Contemporary Topics about Phosphorus in Biology and Materials

Edited by David G. Churchill,
Maja Dutour Sikirić, Božana Čolović
and Helga Füredi Milhofer

Published in London, United Kingdom

IntechOpen

Supporting open minds since 2005

Contemporary Topics about Phosphorus in Biology and Materials
http://dx.doi.org/10.5772/intechopen.80727
Edited by David G. Churchill, Maja Dutour Sikirić, Božana Čolović and Helga Füredi Milhofer

Contributors
Piotr Dobryszycki, Marta Kalka, Anna Zoglowek, Andrzej Ożyhar, Oueslati-Omrani Refka, Ahmed Hichem Hamzaoui, Khattech Ismail, Mohamed Jemal, Pinki Dey, Pawel Kafarski, Drago Skrtic, Diane Bienek, Anthony Giuseppetti, Latefa Sail, Dean Rosenthal, Cynthia Simbulan-Rosenthal, Bonnie Carney, Jeffrey Shupp, Manish Moghe, Anirudh Gaur, Elliott Crooke, Lauren Moffat, Sílvia Gavinho, Manuel Pedro Graça, Sayantan D, Sumona Sanyal Das, David G. Churchill, Woo Hyun Lee, Ariq Abdillah, Jaymee Palma

First published in London, United Kingdom, 2020 by IntechOpen
IntechOpen is the global imprint of INTECHOPEN LIMITED, registered in England and Wales, registration number: 11086078, 7th floor, 10 Lower Thames Street, London, EC3R 6AF, United Kingdom
Printed in Croatia

British Library Cataloguing-in-Publication Data
A catalogue record for this book is available from the British Library

Additional hard and PDF copies can be obtained from orders@intechopen.com

Contemporary Topics about Phosphorus in Biology and Materials
Edited by David G. Churchill, Maja Dutour Sikirić, Božana Čolović and Helga Füredi Milhofer
p. cm.
Print ISBN 978-1-78985-039-0
Online ISBN 978-1-78985-040-6
eBook (PDF) ISBN 978-1-83881-125-9

For EU product safety concerns:
IN TECH d.o.o., Prolaz Marije Krucifikse Kozulić 3, 51000 Rijeka, Croatia, info@intechopen.com or visit our website at intechopen.com.

Meet the editors

David G. Churchill is a full professor in the Department of Chemistry at Korea Advanced Institute of Science and Technology (KAIST). He is an adjunct professor in the KAIST Institute for Health Sciences and serves as Head of the International Faculty Council at KAIST. His current research interests relate to main group synthetic chemistry and dementia research; his publications have recently focused on chemosensing and porphyrinoid chemistry. He has created and (co)organized the biannual International Symposium of Molecular Neurodegenerative Disease Research (ISMNDR) at KAIST (Korea, Daejeon), which was initiated in 2012. He has given seminars worldwide, published more than 120 articles in peer-reviewed journals, and has an H-index of 31. In 2017, he served as an associate visiting professor at IIT Technion in Haifa, Israel.

Maja Dutour Sikirić received her doctorate from the Faculty of Sciences, University of Zagreb. During her studies she has acquired experience and developed an interest in organic - inorganic interactions in crystallization, as related to biomineralization, and physicochemical properties of novel gemini surfactants in solution and solid state. She joined Casali Institute of Applied Science, The Hebrew University of Jerusalem for Postdoctoral studies during which she was working on the development of organic-inorganic nanocomposites, to be used as coatings for metal prosthetic implants. After returning to Ruđer Bošković Institute she has continued her work on crystallization of calcium phosphates, nanocomposite biomaterials as well as using DLS for characterization of nanoparticles in different matrices. She is currently employed as Head of the Laboratory for biocolloids and interface chemistry, Division of Physical Chemistry.

Helga Füredi-Milhofer received PhD from the Faculty of Sciences, University of Zagreb in 1963. Subsequently she was postdoctoral researcher at Case Institute of Technology Cleveland Ohio, USA. In 1956. she joined Ruđer Bošković Institute, where she was the head of Laboratory for Precipitation Processes from 1976 – 1995. In the period 1990-991 she was visiting professor at the Weizman Institute of Science, Rehovot, Israel and from 1991-1992 at Casali Institute of Applied Chemistry, the Hebrew University of Jerusalem, Israel. In 1995 she returned to Casali Institute of Applied Chemistry where she was professor until retirement. Her research interests are i) nucleation, growth and phase transformation of ionic and molecular crystals ii) biological and pathological mineralization iii) 0rganic-inorganic nanocomposites based on amorphous and crystalline calcium phosphates.

Božana Čolović graduated in 2007 and received her PhD degree in 2013, from the Faculty of Technology and Metallurgy, University of Belgrade. Since 2007 she has been employed at Vinča Institute for Nuclear Sciences. Her research is focused on development of biomaterials for application in bone tissue engineering and dentistry: synthesis of ceramics (hydroxyapatite, calcium silicate) and their modification with surface active substances to improve biocompatibility, as well as the deposition of bioactive coatingson the surface of metal implants to improve their mechanical and corrosion properties. So far, she has published 1 chapter in a book, 30 papers in international journals and 10 papers in national journals, presented 21 papers at international conferences and gave 3 invited lectures.

Contents

Preface

Written by experts in the field, this book addresses a diverse set of topics regarding phosphorus chemistry, namely phosphates and closely related chemical systems. Both the natural and the synthetic aspects of phosphates and in some places phosphonates are presented in this book, which is divided into two parts: "Dynamics of Phosphates and Normal Biological Function" and "Phosphates in Biomaterials." The disciplines extend through biochemistry, chemistry, materials science, and engineering. We hope that this contribution serves as both a solid information source as well as a basis for inspiring future discoveries. We are grateful to all contributors and editors and wish them well in their careers and safe travel when conference-going.

David G. Churchill
Molecular Logic Gate Laboratory,
Department of Chemistry,
Korea Advanced Institute of Science and Technology (KAIST),
Daejeon, Republic of Korea

KAIST Institute for Health Science and Technology (KIHST)
(Therapeutic Bioengineering),
Daejeon, Republic of Korea

Maja Dutour Sikirić
Laboratory for Biocolloids and Surface Chemistry,
Ruđer Bošković Institute,
Zagreb, Croatia

Helga Füredi-Milhofer
Casali Institute of Applied Cheistry,
The Hebrew University of Jerusalem,
Jerusalem, Israel

Božana Čolović
Vinča Institute of Nuclear Sciences,
Belgrade, Serbia

Section 1

Introduction

Chapter 1

Introductory Chapter: Phosphorus - Nature's Versatile Pentavalent and Tetrahedral Covalent Building Block and Agent for Energy, Disease and Health

Woohyun Lee, Ariq Abdillah, Jaymee Palma and David G. Churchill

1. Introduction

It is extremely exciting to be involved with editing and discussing the recent developments of phosphorus (P, [*Ne*]$3s^23p^3$), an element that has one main isotope [1]. The element, its compounds, and its reactions (including emergent reactions) will certainly allow for some of "tomorrow's" methods and molecules; this science will help to form an extremely wide swath of emergent technologies. In terms of inorganic chemistry, the element of phosphorus (electronegativity = 2.06) has many future prospects. In the chemistry classroom, we often first discuss allotropes. With well-known diversity known for the allotropes of P that exhibit diverse stability, color, and structural characteristics (atomic number 15, group 15), we are afforded a first look at the possibilities of rich *covalent* chemistry that the element affords; examples of extensive ionic bonding (e.g., Na_3P) are rare. P retains, as always, the chemistry that is extremely dynamic, diverse, and developing.

Phosphorus often takes on certain common forms (**Figure 1**), phosphines, as well as phosphites, and so forth are important classes of ligands in which the free lone pair offers a relatively π-acidic dative bonding interaction (in comparison to amines)

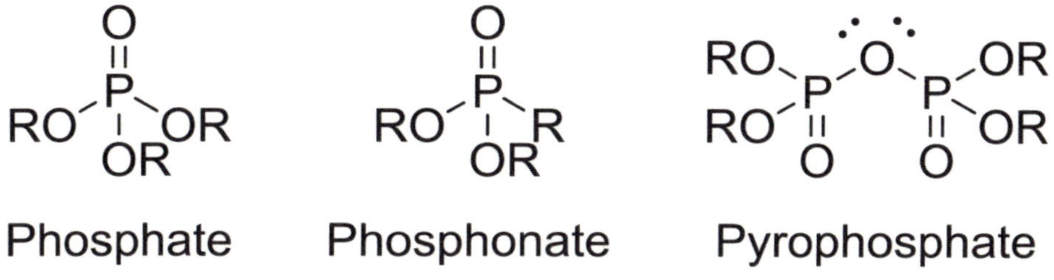

Figure 1.
Structures of simple selected oxygenated phosphorus (V) compounds: phosphate, phosphonate, and pyrophosphate.

with a range of metals [2, 3]. Reports and studies of molecules with the (V) oxides/hydroxides, etc., are rife in health; these species are essential for life as we know it; life sciences are full of such reports (**Table 1**); geochemistry too involves phosphates as expected, an initial literature search in the ISI web of science on the term "phosphate" gives 396,702 hits (since 1970) [4]. While not always the case, phosphorus, and especially its oxo chemistry, is found as a rule to be tetrahedral (pyramidal), and for that matter, pentavalent [5]. While in terms of valence states, there are also valencies of III, however, even when looking in the allotropic forms, we can appreciate the *MLX*(*Z*) covalent bond classification, not the formal oxidation state model [6]. However, the trigonal bipyramidal form is an important transition state intermediate at play and does exist for molecules such as PF_5, which bear certain well defined dynamic processes in solution (see Berry Pseudo rotation) [7]. If we add a fluoride, we can easily formally covert to PF_6^-, which is clearly octahedral and a ubiquitous non coordinating ion (anion). Therefore, as a starting point, we present the information from searches (**Figure 1**). Phosphine (PH_3), for example, is a common parent molecule that is rarely encountered in the lab; as a compound, it has three hydrides and is often considered theoretically when considering phosphine chemistry (see below).

In (bio)materials science terms, tetrahedral allow for bridging and networks and support. Ionic or hybrid materials, crystallization or not, can be encased, or spangled with phosphates; well-ordered materials from a crystalline perspective such as chains, rings, 2-D layers, and 3-D architectures are enabled. These larger molecules are, therefore, very important for materials chemistry. There is also the option for them to be involved with non-well-ordered materials such as glasses, glazes, coatings, etc. In further research, however, we hope that the community can go against the *plus ça change, plus c'est le meme chose* notion.

One P–O–P bond in ATP releases 7–9 kcal mol^{-1} of energy upon hydrolytic scission [8]; analogously, phosphorylation of hydroxyl groups requires 6–9 kcal mol^{-1} in Gibbs free energy [9]. There is, of course, much more to the story, especially the evolutionary story. Considering this chemical energy, chemists studying prebiotic chemistry claim that the recruitment of phosphates for energy in biological systems may have been absent; a predecessor system of energy storage was present (see thioesters) [10]. In other words, phosphate centers that are currently functional serve as a thioester-like center. In light of it being 2020, 1 year since the 450th anniversary of the discovery of the element, and considerations of this simple "light bearing" main group element having been discovered by H. Brand (1669), the same year when Stradivari built his first violin and when Mt. Aetna had a notable eruption, still offers an extremely important amount of contemporary information for today's research community.

Also, in light of the recent novel coronavirus outbreak and the ongoing COVID pandemic, we mention at least one phosphate-based molecule that has been discussed extensively within the research community in the context of enzymatic inhibition;

Topic	Numerical results of articles (hits)
(Phosphorus or phosphate) and "biochemistry"	2475
(Phosphorus or phosphate) and "organic chemistry"	270
(Phosphorus or phosphate) and "analytical chemistry"	168
(Phosphorus or phosphate) and "physical chemistry"	91
(Phosphorus or phosphate) and "inorganic chemistry"	65

Table 1.
ISI Web of Science *search terms related to chemistry subfields presented in descending number of research articles (hits since 1970) [4].*

other important biomolecules that contain phosphates, etc., are also briefly discussed. Here, in this *e-book*, with nine different contributions from institutes worldwide, we can regard new (bio)chemistry considerations and new materials.

1.1 Selected contemporary research fields, industrial workings, and the economy

The involvement of phosphorus in the chemical sciences is truly extensive. It "runs the gamut" from the theoretical chemical sciences and physics through engineering and including all of biology and the life sciences; this breadth and the current access to search the literature enables researchers in one field to access which is working well, or not, in other scientific (technological) fields. We can behold various cycles. There is what is known as the *phosphorus/phosphate cycle* and the *phosphorus/phosphate economy*; phosphorus is traded on the stock exchange [11, 12]; there is the *phosphate metabolism* in the human body. Moreover, there is a great deal in terms of materials/biomaterials to cover and derive as research; with regard to natural bone, teeth, and tissue (*vide infra*); entire journals are dedicated to these research topics (e.g., *Bone*, *Journal of Bone Oncology*, *The Journal of the American Dental Association*, *Journal of Dental Research*, *Journal of Dentistry*, *Journal of Dentistry for Children*, and *International Journal of Oral and Dental Health*). With a tremendous amount of industrial, agricultural (herbicide), and even military importance (white phosphorus and nerve agents), exploration and utility of this element continue to flourish. Therefore, initial key search terms unveil a lot of research activity as found in **Table 2**; these terms, however, do not necessarily articulate the exact research fields involved; therefore, a secondary search was performed and its results are shown below to give a breakdown by research area (**Table 1**).

2. Agriculture

There exists the phosphate cycle; phosphorus, while a macronutrient for plants and animals is centrally important in biology and required in the food chain (*vide infra*). There are always new ways to explore phosphorus-based agriculture agents in the future, however. Of contemporary concern might be, how a heavily used chemical agent such as *Roundup* can be understood fully in terms of its biochemistry; because of some recent controversy, there have been unsubstantiated allegations while prevalent use of the pesticide is ongoing. In the future, we believe that its biological workings will be confirmed fully in molecular detail; our current scientific position is it is *neither* a cancer causing agent *nor* an agent that creates adverse health effects within reason. This exploration of chemical science that is strongly coupled with biology will continue to play out in other examples too.

3. Biochemistry

ATP is an energy storage molecule for life; related molecules also possess the same energy in the anhydride P–O–P bonds; as seen in **Table 1**, considering the number of hits (articles), and looking a little deeper suggesting there may be a lot of research momentum in the field. Furthermore, post-translational modifications (PTM's), for example in the form of phosphorylation, are exceptionally interesting and important; research with this chemistry continues (**Table 2**); the Park laboratory at KAIST for example, has found ways to explore phosphoserine "stop codon" chemistry and to apply this biosynthesis to various problems in molecular biology [13–16]; Additionally, the Lashuel group at EPFL and other researchers (e.g., E. Mandelkow) have been involved with detailed studies of phosphorylation, specifically in the context of neurodegenerative disease research [17–20]. The presence of a phosphorylated serine, threonine, tyrosine, etc., reveals the evident workings of kinases and anticipation of the workings of phosphatases. Among the

Topic (entered search term)	Numerical results (hits)	Notes	People, places and research fields
Phosphorus	207,800	Highly Cited in Field (1198) Hot Papers in Field (27) Open Access (36,011)	—
Phosphate	397,854	Highly Cited in Field (1530) Hot Papers in Field (21) Open Access (91,329)	Centre National de la Recherche Scientifique (CNRS) (10,095) University of California System (8822) Chinese Academy of Sciences (8317) Russian Academy of Sciences (5055) National Institutes of Health (NIH, USA) (3995)
Phosphate and material	41,569	Highly Cited in Field (292) Hot Papers in Field (6) Open Access (7409)	—
Phosphate and NMR	14,062	—	—
*Phosphon**	45,743	—	—
Phosphonate	12,975	—	—
Roundup	2530	—	—
Phosph and virus*	36,004	—	—
*Phosphorus and organometal**	1045	—	—
Phosphate and inhibitor	24,237	—	—
Phosphate and inhibitor and enzyme	7967	Highly Cited in Field (27) Open Access (2720)	Centre National de la Recherche Scientifique (CNRS) (248) University of California System (218) National Institutes of Health (NIH USA) (143) University Of Texas System (135) Russian Academy of Sciences (119)
Phosphate and dementia	346	Highly Cited in Field (3) Open Access (120)	Neurosciences (137) Clinical Neurology (57) Biochemistry Molecular Biology (55) Pharmacology Pharmacy (34) Geriatrics Gerontology (22)
Phosphoryl and dementia*	3897	Highly Cited in Field (65) Hot Papers in Field (2) Open Access (1794)	Blennow K (166) Zetterberg H (128) Trojanowski JQ (106) Scheltens P (68) Lee VMY (66)

The asterisk (*) serves to truncate the search term and thus broaden the search.

Table 2.
ISI Web of Science *search on terms related to phosphorus and the number of articles (hits since 1970).*

other recent findings of import involves biochemistry involving pyrophosphate as a type of PTM, a related phosphate-based species (**Figure 1**) [21–24].

4. Enzyme inhibitors and the COVID outbreak

In recent months, there has been a great emphasis on the possibility of small molecules that can serve as emergent molecular treatments for the novel corona virus (COVID-19). While vaccine research continues, some potential pharmaceutical treatments have recently come to light. Unfortunately, many molecules have been "bandied about" in the news and social media and overly discussed in an unprofessional way by non-experts, and dangerously so, in some cases. However, there has been special and recent attention paid to specific novel drugs that, while still in their early days, are strongly considered by the research community and officials at NIH and elsewhere. Remdesivir, for example (Gilead Sciences, Inc.), was involved in a recently completed trial regarding COVID; some promise was evident (**Figure 2**). The molecule is provided here because it incorporates a phosphate and a description of how it works is important to be illustrated for the readership of our book.

4.1 Remdesivir

4.1.1 How It Works

Remdesivir (**Figure 2**) is a known synthetic molecule developed by Gilead Sciences, USSCD, and USAMRIID. Remdesivir's working chemical mechanism is that of an antiviral nature; it is nucleoside-modified based drug that without such modification has poor cell permeability. The process of transcription of RNA polymerase is blocked in a patient (human subject). Specifically, although Remdesivir still has its 3′hydroxyl site that can form phosphodiester bond, the steric 1′-position containing a cyano group (–CN) prevents any further elongation of the polymerase polymer. Thus, the production of the polymer is halted with this monophosphate, ester, and phosphoramidate-based drug [25]. It is also important to consider the prescribed human dosage currently being considered. Administration of remdesivir is through intravenous injection (IV); the loading dose on the first day is in the amount of 200 mg for an adult with weight adjustment required; following this is maintenance for up to 10 days with a daily dosage amount being 100 mg for an adult.

When considering living cells, (eukaryotic) cells, the known intended mode of action of Remdesivir can be discussed. Remdesivir undergoes hydrolysis at

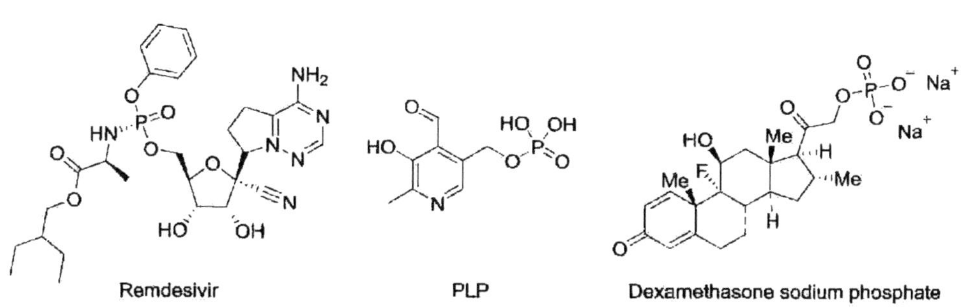

Figure 2.
(Left) Structure of the synthetic Remdesivir proposed as a treatment for the continuing corona outbreak (COVID-19). (Right) The structure of PLP, a naturally occurring small molecule and aldehyde that is phosphorylated.

its aryloxy phosphoramidite; this affords a cyclic anhydride; hydrolysis of such groups is readily obtained through the action of esterases in biology. This cyclic anhydride functionality undergoes reaction with water easily to give the analine. Then, a subsequent hydrolysis event allows for the P–N bond to be cleaved; here, a *phosphoramidase*-type enzyme is at play.

5. Bone mineralization

Bone is crucial and has a multi-important role for any organism. In the next two paragraphs, we want to discuss bone because of its overwhelming importance and the complex process involved in its formation.

Bone is able to help other body organs; also, the process of hemostasis, and the glands of the endocrine system are assisted. The materials comprising bone can be considered in different molecular classifications; there is (i) hydroxyapatite ($Ca_5(PO_4)_3OH$), (ii) collagen, and (iii) water, consistent by weight (65%, 20–30%, 10%, respectively). Collagen is a three-stranded wrapped biomolecule (triple helical structure); bone consists mainly (90%) of collagen type I. Water exists within the hard materials as almost a solvent of crystallization, as well as free water molecules. The process of bone mineralization, therefore, will relate to the ability of the water molecules to move freely during this process. The authors of this text do go into more details. Stiffness relates to the greater percentage of mineral content of bone versus amount of water. The implementation of carbonate is also important; in the beginning of bone formation, mineral in bone consists of both calcium carbonate and calcium phosphate. Through a so called “maturation process,” the carbonate fraction is lowered. The entire process of bone growth has molecular aspects that are still not fully appreciated. The growth of bone involves “medallion” shaped crystals that grow both in the same direction as, and perpendicular to, the fibrils of collagen. The role of carbonate and details of collagen and the microfibrils and related proteins will be discussed in this process and is also explored in this book.

Therefore, phosphate and hydroxyapatite do not come together to compose bone alone; rather it is biosynthesized; the process involves a lot of factors, enzymes and cofactors. For these reasons, research of bone continues and relates significantly to materials chemistry in contemporary research.

6. Synthetic organoelement chemistry

Representative element and organometallic chemistry are rich areas of research. Organometallic chemistry as a discipline is broad; it often encompasses organoelement chemistry and the various reactions by which, e.g., small molecules can be brought into existence. Much emphasis deals with the chemistry industry as well. We can refer to a text such as *Advanced Inorganic Chemistry* 5th edition [2], and *Chemistry of the Elements* [3]. There are also dedicated texts and journals to this main group chemistry topic, journals such as *Phosphorus Sulfur* and *Silicon and the Related Elements* (Publisher: Taylor & Francis). The journal *Organometallics* (ACS) is also an excellent source of new research in organoelement chemistry as well as the journal *Main Group Chemistry*. A couple of recent organoelement and organometallic examples are of note are provided as illustrative examples: the Cummins (MIT) *phosphatetrahedrane* (tri-tert-butyl phosphatetrahedrane, $P(C^tBu)_3$), for example, involves a central unit that resembles a saturated cyclopropane type ring bearing an apical *trivalent* P (**Figure 3**) [5]. Bond strain is apparent. The small organometallic systems also help us to set up a discussion of symmetry and electronics as well as macromolecular possibilities.

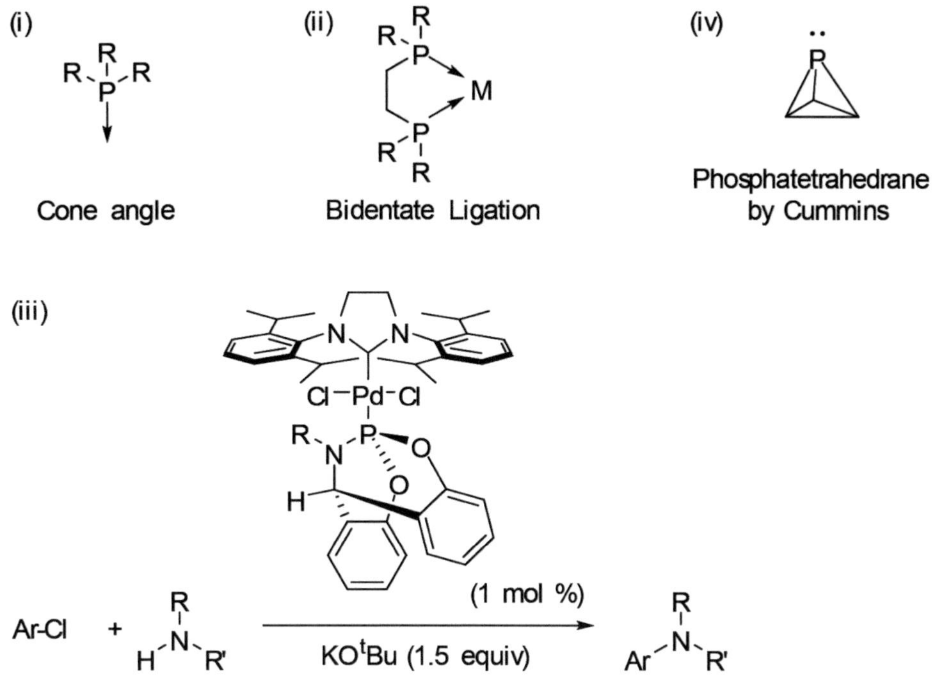

Figure 3.
(i) Monophosphine form that inspired the Tolman cone angle analysis [26]; (ii) the well-known bidentate ligation from diphosphines (e.g., 1,2-bis(diphenylphosphino)ethane); (iii) the recent phosphatetrahedrane *core brought synthetically into reality; and (iv) the important mixed bismonodentate ligation in a recent Pd structure by Kim et al. [27].*

Examples, such as the *phosphatetrahedrane*, continue to show us how different chemistries can be blended to afford creative main group organometallic examples that can factor into current catalysis, also be pursued in current and emergent applications.

7. Catalysis

Phosphorus chemistry afforded phosphine ligands; the corpus of monophosphine ligand research eventually lent itself to a steric bulk analysis known as "cone angle" measurement analysis (Tolman cone angle) (**Figure 3**). Lewis acidic metal centers are often the sites for homogeneous catalysis. The phosphine ligand provides a single soft donor atom; the ligand can be styled frequently as monodentate and bidentate. Other common ligand options are phosphites and phosphoramides and combinations. Phosphines, historically ubiquitous as the *de facto* go-to ancillary ligand, have given way to other ligand families such as carbenes (i.e., *N*-heterocyclic carbenes, NHCs); but there is some very intriguing research on phosphorus-based ligands. A recent intriguing example from KAIST involved a monodentate phosphine analog ligand paired trans with a single N-heterocyclic carbene (NHC) ligand (**Figure 3**) [27]. There is also an important issue of how phosphorus, reduced or oxidized phosphorus is involved in catalyst poisoning or inhibition [28, 29]. Therefore, having P(III) in substrates such as inhibitors, etc., in biochemistry is also an important issue.

8. Characterization

NMR spectroscopy dominates solution characterization in the chemical sciences; it is especially convenient when a (soluble) molecular sample contains P can be subjected to ^{31}P NMR spectroscopy (^{31}P isotope is 100% abundant). To think how spectroscopic characterization can be more extensively organized, phosphorus-31 is uniquely situated in this research field; we can briefly think about both heavier and lighter congeners of group 15 in which the situation is much less convenient. For antimony, the ^{121}Sb isotope and its chemistry can be studied with Mossbauer spectroscopy. Nitrogen 15 (^{14}N is naturally abundant) needs to be introduced into biology as an enriched component (^{15}N-labeled nutrients, etc.) to afford 2-D NMR studies; beautiful solution NMR spectroscopy characterization using ^{1}H-^{15}N HSQC experiments are found in many biochemistry and protein-related articles [30].

9. Materials and biomaterials

As shown in **Table 2**, there are a lot of chemistry topics we could delve into here; there are health-related and biomaterial-related issues with phosphorus. Physical properties such as NMR spectroscopy (as just discussed), thermochemistry, *pKa*, solubility, thermogravimetric analysis (TGA) etc., purification also comes into play. Therefore, this general tetrahedral (sometimes pyramidal) nature and the pentavalent nature serve it well; these boundaries do bear some flexibility. Further, the nature of the intermediates, and some stable compounds keep our mind occupied with how P centers be employed in emergent applications by geometrically tweaking what is apparently preferable based on knowledge of the most stable confirmations of P in a given molecule.

10. Contributions for this *Intechopen* book project

In this work supported by *Intechopen*, we feature the recent advances in phosphates and materials and related studies.

- There is a contribution by Sayantan et al. (CHRIST University, India) discussing about the effect of phosphorus on plants and agriculture. They discuss phosphate in the environment. The types of P-containing species dissolved readily observed in nature. He describes P well as an element. Sayantan explains heavy metals such as Cr and As are taken up by plants causing toxicity; phosphorus mitigates too. In fact, the concentration of P and As in root and shoot tissues of *A. viridis* showed a negative correlation [31]. Plant species such as *Pistia stratiotes* (water lettuce) were also introduced. The combined involvement of Cr and As in the context of phosphate (phosphorus in the environment) help frame important chemical remediation issues.

- The chapter by Kalka et al. (Wrocław University) deals with the proteins that enable biomineralization which involve phosphates [32]. These key processes in biology help afford composition of bone and teeth (tooth, enamel, dentin, cementum, etc.) (see above section). The authors therefore explain nature's structural building materials and their construction therefore are extremely diverse; this is an especially great reference for those biochemists interested in expanding their research into the biomaterials area.

- The chapter by Bienek (Research Division, American Dental Association Foundation) discusses the application of amorphous calcium phosphate composites (ACP) in dental materials. ACP composites can regenerate tooth material through the release of Ca^{2+} and PO_4^{3-} (HPO_4^{2-}, $H_2PO_4^{-}$) ions, making them good candidates for dental restoratives (Research Division, American Dental Association Foundation) [33].

- The reference by Omrani et al. (National Center For Research in Materials Sciences, Tunisia) discusses glasses, material categorically separate from crystalline solids [34, 35]. The sought values of T_g, as well as the NMR spectroscopy, as well as calorimetry were employed (the value T_g is defined as the glass transition temperature) are listed and discussed.

- The chapter by Rosenthal et al. (Georgetown University) deals with *PolyP*; this amazing species is the polymeric form of phosphate [36]. Consider pyrophosphate being elongated to make an enormous polymeric and relatively energetic chain (chemical energy in the P–O–P anhydride bond); these molecules exist and are fascinating. They are prebiotic molecules, and of course, vary in length. They would be naturally quite difficult to study, even by ^{31}P NMR. There is an excellent discussion of the concentration levels of these *PolyP* species in eukaryotes. From studies of rats and mice, PolyP concentration appears high in brain tissue (~100 μM) and PolyP is known for regulation of protein translation, and normal osmolarity [37].

- The chapter by Kafarski (Wrocław University) involves a discussion of the biological use of phosphonates; these are phosphates groups in which there is one carbon attachment (see **Figure 1**) [38]. Biogeochemistry and phosphorus cycling is discussed. Peptides that contain phosphonate groups, specifically phosphonopeptides which can be used as antibiotics and their mechanism of action, are shown. The authors show the biosynthesis of phosphonates. Structures of fosmidomycin analogs are also shown and discussed. Phosphonates can help form methane [38].

- The contribution by Graça et al. (University of Aveiro, Portugal) deals with materials, and calcium phosphate bone cements, abbreviated as CPC's. These species will involve solution chemistry, a dissolution and precipitation (nucleation) processes; this will require RT or physiological temperature [39]. There is microporosity and considerations of optimum concentrations of additives—quite a long discussion of apatite cements and brushite cements is provided. Applications such as drug delivery were discussed; there is a nice scheme showing the interrelated papers. There are also sections for applications dealing with growth factor addition, ion addition and 3D printing is also included.

- The article by Dey et al. (National Institute of Technology, India) deals with bone materialization. As we began to explain above, the complex bone matrix has important roles for bone mineralization [40]. To help make deposition of mineral and for it to become ordered in patterns, nucleation is necessary and made possible by this apparatus. Beside the physiological mineralization, about which the normal mechanisms are described below, pathological bone mineralization also occurs; disabling conditions such as *osteomalacia* and *osteodystropy*. Further, the concept of mineralization has been brought to many applications such polymer science, bio-ceramics, bio-active glasses. While research is still ongoing, there are two models to explain the mechanism of bone mineralization [40] (i) collagen-moderated mineralization and (ii) matrix-vesicle moderated mineralization.

    - In the first proposed mechanism, collagen provides a molecular template for the nucleating mineral that is accumulating. To create nucleation, calcium triphosphate formation emerges from acidic polypeptide pre-nucleation. This molecule is negatively charged; because of its charge, calcium triphosphate interacts with the positively charged regions of the structural collagen template and transforms into a solid phase.

- In the second proposed mechanism, this involves carbonic anhydrase. This allows for the intervesicular environment to become more basic (high pH). The primary crystal material therefore becomes stable. After the hydroxyapatite forms, it becomes in contact with the extracellular fluid vesicle breaks; next, the calcium ion and phosphate ion concentration allow for proliferation of the new hydroxyapatite crystal. What caused the rupture of the vesicle needs further investigation; it is believed that proteases and phospholipases are involved. Mineral transfer is enabled by metalloproteinases. Channeling for mineral transfer is made through the help of collagen types II and X and the outside surface of the matrix vesicle.

- Catalysis involves instances where there is a decrease in efficiency upon contact with unwanted additives (think thermodynamic sink); therefore, inhibition or poisoning can occur [28, 29]. Inhibition, however, might be a good prospect when the reaction is that which deals with metal oxidation (corrosion). With steel (girders and reinforcement) and cement-based construction [41] being what they are in the world today, and the need for industrial reinforcement, Dr. Sail (Abou Bekr Belkaid University) studied the effects of different phosphates; albeit closely related phosphates, were used as corrosion inhibitors [42]. Gravimetric tests were performed; corrosion rates and inhibitor efficiencies were obtained by direct measurements. Assays involving steel corrosion inhibition were measured at variable temperature. While these projects are outstanding and essential to consider in engineering, on a chemical level, the exact mechanistic details at the molecular level are eventually necessary to achieve a full understanding. As part of the proposed mechanism, chlorides were discussed to migrate through pores in the concrete material to reach the steel surface and act to promote corrosion. Steel simply corrodes faster in salt water (sea water) (electrolytes).

11. Research of the Churchill lab at KAIST

As a laboratory involved with synthesis and since ca. 2009 trying to carry forward a big picture of neurodegenerative disease research and chemosensing [44], we have had a chance to publish research that deals with phosphorus chemistry. We have been involved in work that deals with phosphine ligands [45, 46], phosphate metal binding [47–49] (**Figure 4**), phosphyl cleavages in chemical sensor constructs,

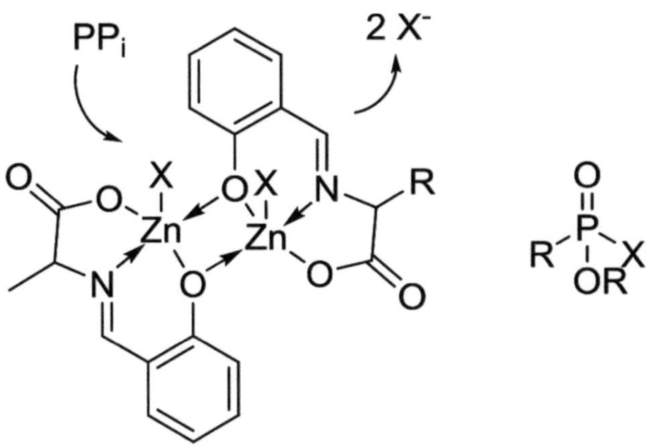

Figure 4.
Dizinc Schiff base dimer synthesized, characterized, and tested in the context of PPi *detection [43].*

and also reviewed the organophosphonate pesticides and CWA literature among other work [24]. Research also extended to theoretical studies as well and porphyrin derivatives. This research involved the themes of characterization, molecular sensing, molecular logic gating and neurodegenerative disease research aspects, and live cell imaging. Recently, while on sabbatical at IIT Technion (Israel), Churchill with Professor Zeev Gross was involved with the synthesis, derivatization and characterization of phosphorus-centered porphyrinoids (corroles); these were studied as important analogs and isomers of beta-position trifluoromethyl substitution [49].

12. Further comments and outlook

There is a relative wide "storm" of other findings related to phosphorus chemistry every year which help expands currently existing research; they are interconnected with other aspects of elements in biology as seen in both **Tables 1** and **2**. We have touched on just a few topics that we currently find interesting, but there is a host of others.

12.1 Molecular sensing

Phosphorus can be a constituent of an analyte, probe, or construct, as well as a connecting or bridging chemical motif (MS-325, MRI contrast agent). In many cases, phosphate has been used as the analyte for such research. There is no perfect sensor; our interests are reminiscent and seated in the goals spelled out by the Sessler et al. review [50, 51] and the Yoon et al. review [50, 51]; our *Molecular Logic Gate Laboratory* yearns for progress relating to synthetic systems that are elaborate optical imaging driven by ATP and GTP and alternations of optical or chemical outputs. Our vision of the chemistry is to address both usable components of analytes themselves as endogenous targets, as well as the novel synthetic pursuits of preparing sensing components.

12.2 Ligands for catalysis

Phosphorus is a reliable donor atom in a range of ligands. Next would be their use as ligands: Phosphines have enjoyed decades as the mainstay ligand type in inorganic chemistry (homogenous catalysis) these have given way to new types of ligation systems such as those with redox action involvement; non-innocent ligand platforms continue to be considered. These and other efforts require the continued attention of synthetic chemists (organometallic chemistry etc.)

12.3 NMR spectroscopy

The solution chemistry is easily accounted for by NMR spectroscopy, therefore, the future is bright for NMR spectroscopy in small molecule work, protein chemistry, materials, electronics applications, etc., that contain phosphorus in one form or another (see above).

12.4 "Roundup" herbicide

The agent glyphosate has the trivial commercial name "roundup" and is an extremely heavily used and a very successful herbicide. In the fullness of time, efforts in biological science will be able to establish any and all positive or negative effects with respect to herbicides, some of which appear to be under some recent scrutiny. We feel that "roundup" will be fully "acquitted" in the near future.

Future research and further development into mild but effective, pesticides and agents are important and surely will be conducted in the future. Among other biological science angles, pesticides, and herbicides that are phosphorus-based, there is the connection to health and, e.g., immune system which can be investigated; this area will be of urgent interest to pursue.

12.5 Prebiotic chemistry

As mentioned in the introduction, the chemistry of phosphates is sometimes theorized through either prebiotic chemistry or future chemical research as a replacement to the chemistry of thioesters; these (RC(O)SR') groups may have been in operation as forerunners to the present-day phosphates. How phosphates (ATP and GTP and others) were brought in to replace this alleged previous ancient chemistry is fascinating; research will help shine light on how life comes into being, and how this can relate to the possibility of extraterrestrial life; and what could shape future life extrapolating forward. The extremely interesting topics of polyphosphates, which are also prebiotic, can also be investigated.

12.6 Phosphorus(III) and switchable chiral centers

Recent research by the Feringa Laboratory for example, involves phosphorus species that are involved with chiral switching. In these cases, there is a "trivalent" phosphorus at play. Please see examples of this recent work [52]. How light can be used as an external stimulus and allow for switching of multiple chiral molecular logic gating centers is an important issue where the scientific community can establish certain "world records": (i) most number of chiral centers, (ii) weakest effective wavelength, (iii) most convenient multiplexing, etc.

13. Conclusion

In conclusion, we have used this introductory chapter, to primarily introduce the authors' nine focused chapters. We have also tried to create optimism and describe some semblance of an interconnected foundation for future research; research areas and their connectivity and some examples were laid out. Perceived areas of activity, enthusiasm, and hope for phosphate chemistry, including organoelement chemistry, was brought forth. While we are excited and proud to introduce the topics shown here, from our own laboratory, we delved into some aspects of organoelement chemistry of phosphorus and phosphates. Through the contributions by our authors from Algeria, India, Poland, Portugal, Tunisia, and the USA, and we see how the literature of P-containing molecules continues to expand itself [53, 54]. As reflected from our own experience and publication trajectory, too, here we are looking at the oxidized P(V) chemistry almost exclusively. We introduced each chapter's emphasis. Phosphorus offers an extremely wide palate of chemistry and from a prebiotic point of view too; it, despite tetrahedral/pentavalent disposition, could still be thought of as a visionary futuristic element with research that is "dynamic and interdisciplinary." In some circles, phosphates are best known as being involved on storing energy in the P–O bond (9 kcal mol^{-1}). Because of its affinity to O, yet covalent, phosphates, feature as a macronutrient element for health. There are main group research findings all the time. Regarding future collaboration there is an interesting story about the collaboration, in dentistry and bone research regarding the compound etidronic acid, 1,1-diphosphonic acid (EHDP), a collaboration that was fueled by a coffee meeting [55, 56]. An interesting story about scientists serendipitously initiating work together like this is to help unlock more important findings and utility about the element in the future. Phosphorus is, and for that matter phosphates are, chemical species for life on planet Earth and beyond.

Acknowledgements

J. P. acknowledges the KAIST Winter/Spring semester URP Program (2020). We acknowledge the outstanding help of the staff from *Intechopen* (Marijana Francetic) and coeditorship with Professor Maja Dutour Sikirić, Dr. Božana Čolović, and Dr. Helga Füredi Milhofer. We thank the author contributors for their patience. Dr. David G. Churchill, acknowledges the insights of previous group members (PhD students and postdocs) who have been involved in this area of research including Kibong Kim (Korea Institute of S&T Evaluation and Planning, KISTEP), Dr. Taehong Jun, (Samsung), Dr. Youngsam Kim (Korea Institute of Science and Technology, KIST Europe), Dr. Yoonjeong Jang (South Korea CBRN Defense Research Institute), Dr. Tesla Yudhistira (Université de Strasbourg), Dr. Olga G. Tsay (Samsung), Professor Yonghwang Ha (Jungwon University), Professor Snehadrinarayan Khatua (North-Eastern Hill University), Professor Sudesh T. Manjare (University of Mumbai), and Dr. Sandip V. Mulay (Korea Research Institute of Chemical Technology, KRICT).

Conflicts of interest

DGC currently has no commercial interests in the manufacture, use, etc., of Roundup or other agrochemicals or related agents, or in concrete, its production, and related chemical additives.

Notes

In memory of: Dr. Helga Füredi Milhofer.

Abbreviations

Ar	arsenic
COVID	corona virus diseases
Cr	chromium
DNA	deoxyribonucleic acid
EHDP	etidronic acid, 1,1-diphosphonic acid
EPFL	Ecole Polytechnique Fédérale de Lausanne
ISI	Institute for Scientific Information
Na	sodium
Ne	neon
NMR	nuclear magnetic resonance
PLP	pyridoxal 5'-phosphate
PolyP	polyphosphate
PTM	posttranslational modifications
RNA	ribonucleic acid
SN2	nucleophilic substitution second order

Author details

Woohyun Lee[1], Ariq Abdillah[1], Jaymee Palma[1] and David G. Churchill[1,2]*

1 Molecular Logic Gate Laboratory, Department of Chemistry, Korea Advanced Institute of Science and Technology (KAIST), Daejeon, Republic of Korea

2 KAIST Institute for Health Science and Technology (KIHST) (Therapeutic Bioengineering), Daejeon, Republic of Korea

*Address all correspondence to: david.churchill.korea@gmail.com; dchurchill@kaist.ac.kr

References

[1] Kim K, Churchill DG. Phosphorus: Radionuclides. Encyclopedia of Inorganic and Bioinorganic Chemistry. 2011. DOI: 10.1002/9781119951438.eibc0418. Available from: https://onlinelibrary.wiley.com/doi/abs/10.1002/9781119951438.eibc0418

[2] Cotton FA et al. Advanced Inorganic Chemistry. 1980

[3] Greenwood NN, Earnshaw A. Chemistry of the Elements. 2nd ed. 1997

[4] ISI Web of Science. Clarivate Analytics, 2020. Clarivate (Search Engine Web-Site)

[5] Riu ML et al. Isolation of an elusive phosphatetrahedrane. Science Advances. 2020;**6**:eaaz3168

[6] Green MLH, Parkin G. Application of the covalent bond classification method for the teaching of inorganic chemistry. Journal of Chemical Education. 2014;**91**(6):807-816

[7] Russegger P, Brickmann J. Pseudorotation of trigonal bipyramidal molecules: Berry rotation contra "turnstile" rotation in PF_5. Chemical Physics Letters. 1975;**30**:276-278

[8] Rosing J et al. The value of G degrees for the hydrolysis of ATP. Biochimica et Biophysica Acta. 1972;**267**:275-290

[9] Ni F et al. N-phosphoryl amino acid models for P-N bonds in prebiotic chemical evolution. Science China. Chemistry. 2015;**58**:374-382

[10] Bracher PJ et al. The relative rates of thiol-thioester exchange and hydrolysis for alkyl and aryl thioalkanoates in water. Origins of Life and Evolution of the Biosphere. 2011;**41**:399-412

[11] Weber O et al. Trade and finance as cross-cutting issues in the global phosphate and fertilizer market. In: Sustainable Phosphorus Management. Springer; 2014. pp. 275-299

[12] Nesme T, Metson GS, Bennett EM. Global phosphorus flows through agricultural trade. Global Environmental Change. 2018;**50**:133-141

[13] Yang A et al. A chemical biology route to site-specific authentic protein modifications. Science. 2016;**354**:623-626

[14] Ha Y et al. Facile "stop-codon" method reveals elevated neuronal toxicity by discrete S87p-α-synuclein oligomers. Biochemical and Biophysical Research Communications. 2014;**443**:1085-1091

[15] Lee S et al. A facile strategy for selective incorporation of phosphoserine into histones. Angewandte Chemie, International Edition. 2013;**52**:5771-5775

[16] Park H-S et al. Expanding the genetic code of *Escherichia coli* with phosphoserine. Science. 2011;**333**:1151-1154

[17] Haj-Yahya M et al. Site-specific hyperphosphorylaion inhibits, rather than promotes, Tau fibrillization, seeding capacity, and its microtubule binding. Angewandte Chemie, International Edition. 2019;**59**:4059-4067

[18] Cariulo C et al. Ultrasensitive quantitative measurement of Huntingtin phosphorylation at residue S13. Biochemical and Biophysical Research Communications. 2020;**521**:549-554

[19] Yalinca H et al. The role of post-translational modifications on the energy landscape of Huntingtin N-terminus. Frontiers in Molecular Biosciences. 2019;**6**:95

[20] Cariulo C et al. Phospho-S129 alpha-synuclein is present in human plasma but not in cerebrospinal fluid as determined by an ultrasensitive immunoassay. Frontiers in Neuroscience. 2019;**13**:889

[21] Zhu J et al. Two bifunctional inositol pyrophosphate kinases/phosphatases control plant phosphate homeostasis. eLife. 2019;**8**:e43582

[22] Wu M, Chong LS, Perlman DH, Resnick AC, Fiedler D. Inositol polyphosphates intersect with signaling and metabolic networks via two distinct mechanisms. PNAS. 2016;**113**:E6757-E6765

[23] Brown NW, Marmelstein AM, Fiedler D. Chemical tools for interrogating inositol pyrophosphate structure and function. Chemical Society Reviews. 2016;**45**:6311-6326

[24] Kim K, Tsay OG, Atwood DA, Churchill DG. Destruction and detection of chemical warfare agents. Chemical Reviews. 2011;**111**:5345-5403

[25] Eastman RT et al. Remdesivir: A review of its discovery and development leading to emergency use authorization for treatment of COVID-19. ACS Central Science. 2020;**6**:672-683

[26] Tolman CA. Steric effects of phosphorus ligands in organometallic chemistry and homogeneous catalysis. Chemical Reviews. 1977;**77**:313-348

[27] Kim M, Shin T, Lee A, Kim H. Synergistic ligand effect between N-heterocyclic carbene (NHC) and bicyclic phosphoramidite (Briphos) ligands in Pd-catalyzed amination. Organometallics. 2018;**37**:3253-3258

[28] Kirchner BAK. The poisoning of noble metal catalysts by phosphorus compounds—I. Chemical processes, mechanisms and changes in the catalyst. Chemical Engineering Science. 1980;**35**:2089-2091

[29] Rokosz MJ et al. Characterization of phosphorus-poisoned automotive exhaust catalysts. Applied Catalysis. B, Environmental. 2001;**33**:205-215

[30] Matte A et al. Preparation and characterization of bacterial protein complexes for structural analysis. Advances in Protein Chemistry and Structural Biology. 2009;**76**:1-42

[31] Sayantan D et al. Phosphorus: A boon or curse for the environment? In: Contemporary Topics about Phosphorus in Biology and Materials. 2020

[32] Kalka M et al. Proteins in calcium phosphate biomineralization. In: Contemporary Topics about Phosphorus in Biology and Materials. 2019

[33] Bienek DR et al. Amorphous calcium phosphate as bioactive filler in polymeric dental composites. In: Contemporary Topics about Phosphorus in Biology and Materials. 2019

[34] Omrani RO et al. Structural and calorimetric studies of zinc, magnesium and manganese based phosphate and phosphate-silicate glasses. In: Contemporary Topics about Phosphorus in Biology and Materials. 2019

[35] Fulchiron R, Belyamani I, Otaigbe JU, Bounor-Legaré V. A simple method for tuning the glass transition process in inorganic phosphate glasses. Scientific Reports. 2015;**5**:8369

[36] Simbulan-Rosenthal CM et al. Inorganic polyphosphates are important for cell survival and motility of human skin keratinocytes and play a role in wound healing. In: Contemporary Topics about Phosphorus in Biology and Materials. 2019

[37] Xie L, Jakob UJ. Inorganic polyphosphate, a multifunctional polyanionic protein scaffold. Journal of Biological Chemistry. 2019;**294**:2180-2190

[38] Kafarski P. Phosphonates: Their natural occurrence and physiological role. In: Contemporary Topics about Phosphorus in Biology and Materials. 2019

[39] Fernandes MP et al. Calcium phosphate cements in tissue engineering. In: Contemporary Topics about Phosphorus in Biology and Materials. 2020

[40] Dey P. Bone mineralization. In: Contemporary Topics about Phosphorus in Biology and Materials. 2019

[41] Wagh AS. Chemically Bonded Phosphate Ceramics. 2nd ed. 2016. pp. 17-34

[42] Available from: https://www.intechopen.com/online–first/temperature–influence–on–inhibitory–efficiency–of–three–phosphate–inhibitors–by–mass–loss

[43] Khatua S, Choi SH, Lee J, Kim K, Do Y, Churchill DG. Aqueous fluorometric and colorimetric sensing of phosphate ions by a fluorescent dinuclear zinc complex. Inorganic Chemistry. 2009;**48**:2993-2999

[44] Biannual Conference: International Symposium for Molecular Neurodegenerative Disease Research (ISMNDR). Available from: https://sites.google.com/view/ismndr4/ismndr-4

[45] Kang H et al. The synthesis and characterization of $Re_3(\mu\text{-}H)_3(CO)_{9-n}(PMe_3)_n(\mu_3\text{-}\eta^2{:}\eta^2{:}\eta^2\text{-}C_{60})$ (n=2,3) complexes. Journal of Organometallic Chemistry. 2005;**690**:4704-4711

[46] Park BK et al. Ortho phosphorylation of PPh_3 to give a diphosphine and formation of a "butterfly" structure on a tetrairidium framework. Organometallics. 2005;**24**:675-679

[47] Tsay OG et al. Novel reversible Zn^{2+}-assisted biological phosphate "turn-on" probing through stable aryl-hydrazone salicylaldimine conjugation that attenuates ligand hydrolysis. Inorganic Chemistry. 2013;**52**:10052-10061

[48] Kim K, Ha Y, Kaufman L, Churchill DG. Labile zinc-assisted biological phosphate chemosensing and related molecular logic gating interpretations. Inorganic Chemistry. 2012;**51**:928-938

[49] Zhan X et al. Maximizing property tuning of phosphorus corrole photocatalysts through a trifluoromethylation approach. Inorganic Chemistry. 2019;**58**:6184-6198

[50] Hargrove AE, Nieto S, Zhang T, Sessler JL, Anslyn EV. Artificial receptors for the recognition of phosphorylated molecules. Chemical Reviews. 2011;**111**:6603-6782

[51] Lee S, Yuen KKY, Jolliffe KA, Yoon J. Fluorescent and colorimetric chemosensors for pyrophosphate. Chemical Society Reviews. 2015;**44**:1749-1762

[52] Pizzolato SF et al. Phosphoramidite-based photoresponsive ligands displaying multifold transfer of chirality in dynamic enantioselective metal catalysis. Nature Catalysis. 2020;**3**:488-496

[53] Vlatković M, Feringa BL, Wezenberg SJ. Angewandte Chemie, International Edition. 2016;**55**:1001-1004

[54] Pizzolato SF, Štacko P, Kistemaker JCM, Leeuwen TV, Feringa BL. Nature Catalysis. 2020

[55] Francis MD, Centner RL. Journal of Chemical Education. 1978;**55**:760

[56] Williams KR. Journal of Chemical Education. 2004;**81**:1406

Section 2

Dynamics of Phosphates and Normal Biological Function

Chapter 2

Phosphorus: A Boon or Curse for the Environment?

D. Sayantan and Sumona Sanyal Das

Abstract

Phosphorus, a limiting nutrient of biosphere, exists as dissolved inorganic phosphorus (DIP), dissolved organic phosphorus (DOP), particulate inorganic phosphorus (PIP) and particulate organic phosphorus (POP) in water of soil as well as ponds, lakes, etc. The only available phosphorus for plants are DIP, while the other forms need to be converted to DIP by the decomposing microorganisms of the soil. The heavy metals (such as arsenic and chromium), which are the menace of both terrestrial and aquatic environment, are taken up by the plants and animals causing toxicity at physiological level. However, the metal (Cr and As) toxicity can be mitigated competitively by phosphorus, since the latter is a structural analogue. Since, phosphorus is an essential nutrient, plants prefer it over Cr or As. At the same time, if excess of phosphorus is applied in the soil in the form of fertilisers, it gets discharged into the water bodies (ponds, lakes, etc.) through agricultural runoff, causing eutrophication followed by harming the health of the water bodies. This can be further mitigated by employing the phenomenon of luxury uptake by the aquatic plants such as *Pistia stratiotes.*

Keywords: arsenic, chromium, eutrophication, luxury uptake, phosphorus

1. Phosphorus: an introduction

Phosphorus (P) (*fos'furus* = light-bearing) is a reactive, non-metallic, multivalent chemical element of the nitrogen family group 15 [Va] of the periodic table, along with arsenic, antimony and bismuth and moscovium [1]. The atomic number of phosphorus is 15, and it has a density of 1.823 g/cm^3 (298 K), an atomic volume of 17 cm^3/mol, an atomic mass of 30.97 g/mol, three main oxidation states (−3, +3 and +5) and one naturally occurring isotope (^{31}P) (http://epathshala.nic.in/QR/books/12Chemistry/7.1.pdf). Phosphorus has 15 protons and there are 17 known isotopes. It is the 11th most abundant element in the earth's crust, consisting approximately of 0.1% by weight [2]. It was discovered and named by Hennig Brand who was a Hamburg merchant and alchemist by profession in 1669 by boiling and evaporating about 1100 l of urine and heating the residue with sand (which yield hot gases and vapour) and condensing it with cold water. The final substance to condense was soft, waxy white solid material which glowed in the dark. This element was named phosphorus [3, 4]. Phosphorus exists in several allotropic forms showing different properties, three of which are white phosphorus, red phosphorus and black phosphorus, which commonly exists [5].

White phosphorus consists of tetrahedral P_4 molecules, in which each atom is bound to the other three atoms by a single bond. This form generally exists (in liquid and gaseous) as tetrahedral P_4 molecules at temperature up to 800°C that change to P_2 molecules at temperature higher than 800°C [6]. White phosphorus is the most reactive and least stable; for this reason it should be stored under water, as it is dangerously reactive in air, and it must be handled with forceps, as contact with the skin may result in severe burns. Due to its high reactivity, white phosphorus is never found as a free element on earth. It is also very poisonous, 50 mg of which constitutes an approximate fatal dose. Exposure to white phosphorus should not exceed 0.1 mg/m^3 (8-h time weighted average—40-h work week) [7]. It is the most volatile and least dense of all allotropes [8]. Because of its dehydrating nature, it has a corrosive property [9].

Red phosphorus is polymeric in structure. It is non-poisonous, odourless and insoluble in water. It can be viewed as a derivative of P_4. It can be produced by heating white phosphorus to around 250°C within the sight of daylight. Freshly prepared, bright red phosphorus is highly reactive and ignites at about 300°C, though it is still more stable than white phosphorus, which ignites at about 30°C.

Black phosphorus is a layered semiconducting material similar in appearance to graphite with numerous uses in optoelectronic, semiconductor and photovoltaic applications. In a two-dimensional form, black phosphorus is known as phosphorene and has similar properties to other 2D semiconductor materials such as graphene. Among other allotropes, this form is the most thermodynamically stable and densest allotropes under ambient temperature and pressure.

2. Sources of phosphorus

Phosphorus is a naturally occurring element that exists in minerals, soil, living organisms and water. Phosphorus occurs in nature as orthophosphate (PO_4^{3-}), principally in the mineral apatite $Ca_5(PO_4)_3$(F, Cl, OH) but also in monazite (Ce, La, Nd, Th) (PO_4, SiO_4) and xenotime YPO_4. It is widely dispersed at trace levels in minerals such as olivine, pyroxene, amphibole and mica and is also present in biological materials such as the bone. Phosphorus is a major element in all plants. In the natural world, phosphorus is never encountered in its elemental form but only as phosphates, which consists of a phosphorus atom bonded to four oxygen atoms. This can exist as the negatively charged phosphate ion (PO_4^{3-}), which is how it occurs in minerals, or as organophosphates in which there are organic molecules attached to one, two or three of the oxygen atoms. Phosphorus exists in many different forms in soil. In general, these sources can be grouped into three general forms: (i) organic phosphorus, (ii) soluble phosphorus and (iii) adsorbed phosphorus. Phosphorus (in the forms hydrogen phosphate [HPO_4^{2-}] and dihydrogen phosphate [$H_2PO_4^-$]) is least mobile in soil. The comprehension of forms of phosphorus will enable to understand mobility of phosphorus in soil and the extent to which phosphorus can move within the environment.

Organic phosphorus account for 15–80% of the phosphorus in soils, the exact amount being dependent upon the nature of the soil and its composition. It is the principal form of phosphorus in the manure of most animals. About two-thirds of the phosphorus in fresh manure is in the organic form. Nucleic acid constitutes 1–5% of total organic phosphorus in soil.

Soluble phosphorus or available inorganic phosphorus include small amounts of organic phosphorus, as well as the orthophosphates, $H_2PO_4^-$ and HPO_4^{2-} (the primary forms of phosphorus), taken up by algae and plants (aquatic and terrestrial). The soluble form accounts for the smallest proportion of the total phosphorus in most of the soils. When fertiliser or manure (both containing mostly soluble phosphorus) is added to soil, the soil's pool of soluble phosphorus increases. With time, soluble phosphorus is transformed slowly to less-soluble forms. The effluents of sewage treatment plants contain mostly soluble phosphorus. Phosphorus is adsorbed by plants in the ionic forms $H_2PO_4^-$ and HPO_4^- [10]. The adsorbed phosphorus does not leach. In fact, it moves very little, even with large amounts of precipitation or irrigation. Attached phosphorus includes labile or loosely bound, and "fixed," or tightly bound, phosphorus compounds.

Agricultural fields lose adsorbed phosphorus in a mechanism to transport phosphorus to water bodies. Phosphorus travel to surface water attached to particles of soil or manure. Phosphorus can also dissolve into runoff water as it passes over the surface of the field. Soil particles strip soluble phosphorus from the water as it moves through the soil profile. Leaching of phosphorus usually is not a significant concern. The concentration of phosphorus in soil leachate is significantly less than surface runoff concentrations [10].

However, special situations can produce higher concentrations of phosphorus in groundwater. The capacity of soil to absorb phosphorus can be overwhelmed on sandy soils or when the water table is close to the soil surface. Also, cracking in soils creates channels allowing surface water to travel directly to groundwater.

Increased use of phosphorus in fertiliser had unsustainable consequences. Since phosphorus remains a finite, diminishing and irreplaceable resource, this is affecting global phosphorus cycles. Phosphorus in rocks is unavailable to organisms, but now increased used of phosphorus in fertilisers has tripled the rate at which biologically available phosphorus enters the ecosystem.

3. Presence of phosphorus in environment

The phosphorus cycle (**Figure 1**) is the process by which phosphorus moves through the lithosphere, hydrosphere and biosphere. Phosphorus being a constituent of nucleic acids as well as cell membranes and phosphate/phosphorylated compounds is a nutrient of major importance for the biosphere. Calcium phosphate is also the primary component of mammalian bones and teeth and is used in a variety of other biological functions. The phosphorus cycle is an extremely slow process, as various weather conditions (e.g. rain and erosion) help to wash the phosphorus

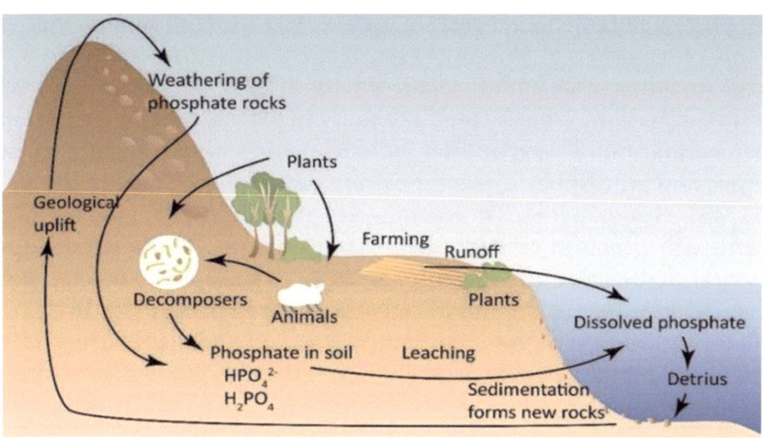

Figure 1.
Phosphorus cycle (image source: Ref. [11]).

found in rocks into the soil. In the soil, the organic matter (e.g. plants and fungi) absorbs the phosphorus to be used for various biological processes [11].

Phosphorus is an important element for all forms of life. As phosphate (PO_4^{3-}), it makes up an important part of the structural framework that holds DNA and RNA together. Phosphates are also a critical component of ATP, the cellular energy carrier, as they serve as an energy release for organisms to use in building proteins or contacting muscles. Like calcium, phosphorus is important to vertebrates; in the human body, 80% of phosphorus is found in teeth and bones [11].

The phosphorus cycle differs from the other major biogeochemical cycles in that it does not include a gas phase, although small amounts of phosphoric acid (H_3PO_4) may make their way into the atmosphere, contributing, in some cases, to acid rain. The water, carbon, nitrogen and sulphur cycles all include at least one phase in which the element is in its gaseous state. Very less amount of phosphorus circulates in the atmosphere because under normal temperatures and pressures condition of the earth, phosphorus and its various compounds are not gases. The largest reservoir of phosphorus is found in the lithosphere [11].

Weathering: since the main source of phosphorus is found in rocks, the first step of the phosphorus cycle involves the extraction of phosphorus from the rocks by weathering. From weathering the released phosphorus is transported to the soil by wind or water as inorganic phosphate. Absorption by plant and animal: inorganic phosphate is absorbed and assimilated by plants. In most soil, the amount of available phosphorus is about 0.01% of the total phosphorus in soil. From the plant, phosphorus moves through the food chain in the organic form. Once the living organisms die, the organically bound P is returned to the environment by the process of degradation by the decomposition. Organically the dead organic matter is acted upon by the phosphatising bacteria to release inorganic phosphorus from bound organic form. Phosphorus is also returned from shallow marine deposits in fish harvest and guano deposits in fish-eating birds and geological uplift. Turnover of organic phosphorus occurs due to phosphatase activity associated with root activity and microbial populations. The precipitation of phosphorus in marine habitats limit primary productivity [1, 11].

4. Role of phosphorus in mitigating arsenic and chromium toxicity from plants

Since phosphorus is a limiting nutrient for terrestrial biological productivity, it commonly plays a key role in net carbon uptake in terrestrial ecosystems. Although the inorganic phosphorus of soil causes harm to the environment in the form of eutrophication, it has some beneficial effects too. The phosphorus present in soils helps in mitigation of heavy metal toxicity in plants. Increasing population; rapid urbanisation; rapidly expanding industrial areas; use of fertilisers, pesticides and manures; and atmospheric decomposition have added lots of pollutants in the soil [12, 13]. Heavy metals like arsenic (As) and chromium (Cr) are highly toxic and cause ill effects at very low concentrations [14]. They are posing a major environmental challenge; since they do not undergo microbial and chemical degradation, they become persistent and bioaccumulative in nature [15, 16]. They affect soil chemistry by altering pH and conductivity of soil and also cause oxidative and physiological toxicity in the plants [17]. Many technologies like electrokinetic (EK) technique [18], electrokinetic-geosynthetic approach [19] and excavation and physical removal of the soil [20] are used to clean heavy metal pollutants. Each remediation technology has its specific benefits and limitations, but in general none of them is cost-effective [21]. Phytoremediation has the benefit of being a relatively low-cost, natural solution to an environmental problem [22]. Some plants are hyperaccumulators of metal. If the hyperaccumulator plants are edible, the roots will take up the heavy metals from the soil, leading to bioaccumulation of heavy metals in the entire plant body, which will ultimately enter into the food chain and pose health hazards to higher organisms. So, in order to stop the entry of heavy metals into the food chain, it is excellent to apply phosphorus in the soil that will alleviate the uptake and accumulation of heavy metals in food crops growing in the soil contaminated with heavy metals like arsenic and chromium (**Figure 2**) [23–25].

Literatures suggest that arsenate and phosphate share a common transport pathway via roots of the higher plants [23, 26, 27]. Since phosphate is a growth-supporting nutrient, plants prefer its uptake over arsenate. With the understanding of the above molecular analogy between arsenate and phosphate, the most cost-effective method for amelioration of arsenic toxicities, at the oxidative and physiological levels, in the agriculture would be the prevention of its entry in the food chain using phosphate amendments. Thus, a study by Sayantan and Shardendu [23] has been designed to determine the effect of arsenate-phosphate amendments on the variation in the physiological and oxidative toxicities in the root and shoot tissues of *Amaranthus viridis L.* By differentiating the total As accumulation at the tissue level, it was found in the present study that the roots of *A. viridis* accumulated up to twofold higher amount of total As than that in shoot. When the PO_4^{3-} was amended along with the AsO_4^{3-} supplies, the total As accumulated in the root and shoot tissues was significantly ($p < 0.001$) reduced by the maximum of 68.18 and 64%, respectively, with respect to control (0 μM PO_4^{3-} amendment). These observations were further supported statistically by obtaining significant ($p < 0.05$) negative correlation coefficient values between the total As accumulation and the total P stored in root and shoot tissues.

Sayantan and Shardendu [24] showed the role of different levels of phosphorus amendments on chromium toxicity in *Raphanus sativus L.* The experiments were done with the design in which five different concentrations of Cr have been taken and at each Cr level; there were five amendments of P concentration. After completion of experiment, results showed chromium accumulation in plant roots

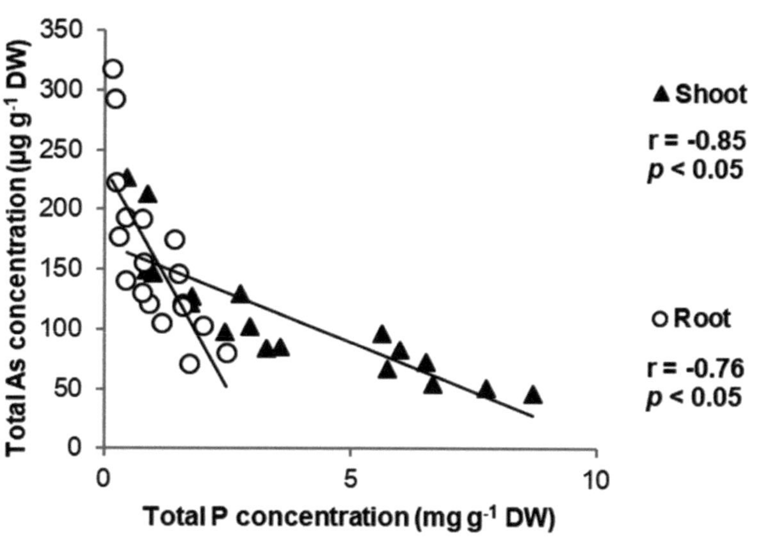

Figure 2.
Relationship between total As and P accumulations in root and shoot tissues of A. viridis, *each with representations of Pearson's correlation coefficient (r-value) and level of significance (p value) (image source: Ref. [23]).*

was dose-dependent and increased as the concentration of Cr supply is enhanced. However, the impact of amendment of phosphorus concentration on Cr accumulation was notable. At each Cr supply, accumulation of Cr decreased as the levels of P increased, showing a reciprocal correlation ($-0.960 \leq r \leq -0.762$) among these two parameters.

Qian et al. [28] showed that chromium could damage the thylakoid structure and impinge on the chloroplast, therefore affecting algal growth. P availability can alleviate Cr toxicity in *C. vulgaris* by decreasing the absorption of Cr and changing the absorption of other metal ions. It is, therefore, necessary to consider the phosphorus availability when the toxicity of metal compounds is evaluated.

It is known that the nutrients present in the soil influence the Cr uptake by *Raphanus sativus* [24]. Sayantan and Shardendu [25] examine the role of Cr toxicity on root and shoot tissues of *Spinacia oleracea* L. (spinach) and its variation when different levels of P were amended with the Cr supply. The path of movement of Cr and P is similar in plants; however, P is preferred over Cr. Thus, P competitively inhibits the Cr uptake. With the phosphorus amendment in the growth medium, accumulation of chromium decreased up to 55% in root and 50% in shoot tissues. These results showed two general observations, i.e. increasing Cr concentrations induced an enhancement in toxicity at both physiological and oxidative levels; and the amendment of P at each Cr supply ameliorated the toxic effects in both kinds of tissues of *S. oleracea*.

Phosphorus is essential to all living beings also; it is one of the main nutrients for animal and plant growth. Phosphorus is used by plants in photosynthesis. When there is less phosphorus during photosynthesis, the production of sugars is restricted. This directly affects the colour of the plant. Phosphorus-deficient plants will not be green but have a purple hue to them. A lack of sugar in the plants will restrict the growth of the plant since sugar is the main source of energy in a plant.

Phosphorus acts as a catalyst. When fertiliser is added to plants, it is often thought of to assist in allowing them to grow quicker and stronger. This is due to the phosphorus in the fertiliser. Phosphorus speeds up plant growth as well as the quality of plant growth.

Phosphorus is also a key component in the structures of life. It is found linking DNA and RNA together and being found in bones of animals. Phosphorus is also involved in ATP that forms during photosynthesis.

Phosphorus is vital to the environment because it allows plant growth that is necessary to keep the ecosystem balanced and flourishing. It allows plants to grow robust to feed the animals that eat them. Healthier plants also allow more oxygen to be released in the air. In manageable amounts, phosphorus can help any ecosystem thrive by providing more food and oxygen. The only measure to check the use of phosphorus for the environment is to reduce the usage of chemical fertiliser and shift towards organic farming. This will reduce eutrophication and, in other words, improve the health of lakes.

5. Eutrophication

When sewage and agricultural runoff containing phosphates or other nutrients enter water bodies, they cause overnutrition, leading to eutrophication. "Eutrophication is an enrichment of water by nutrient salts that causes structural changes to the ecosystem such as: increased production of algae and aquatic plants, depletion of fish species, general deterioration of water quality and other effects that reduce and preclude use" (Organisation for Economic Cooperation and Development). Eutrophication is a major environmental issue as it causes degradation in the quality of water and is one of the major hindrances for achieving the quality objectives established by Designated Best Use Water Quality Criteria by Central Pollution Control Board, India, 2019 [29]. As per the International Lake Environment Committee Foundation, Japan, about 54% of lakes are affected by the phenomenon of eutrophication in Asia ("http://www.eniscuola.net/en/2016/11/03/what-is-eutrophication-causes-effects-and-control/"; "https://www.ilec.or.jp/en/") [30]. Eutrophication occurs naturally over centuries as lakes age and is filled in with sediments [31]. However, anthropogenic activities like discharge of growth-limiting nutrients like phosphorus have accelerated the rate and extent of eutrophication. Any aquatic body starts its life cycle as oligotrophic, i.e. clear body of water. Eutrophication is distinguished by a remarkable increase of algae (very simple, non-flowering aquatic plant) due to the considerable availability of one or more growth factors obligatory for photosynthesis, such as sunlight, carbon dioxide and nutrients (primarily phosphorus) [32]. With the introduction of nutrients through runoff, algae start to grow in an uncontrolled manner. With this growth, increasingly large biomass is formed which is destined to decay. In deep water, pond collects a good amount of organic substance, represented by the algae having reached the end of their life cycle. Eventually, there is algal bloom when the lake becomes marsh or debris. An excessive consumption of oxygen is required by microorganisms to destroy all the dead algae. This created an oxygen-free environment in the lake bottom, anaerobic organisms being responsible for the degradation of the biomass [33]. The microorganisms, decomposing the organic substance in the absence of oxygen-free compounds that are toxic, such as ammonia and hydrogen sulphide (H_2S), were formed. The absence of oxygen reduces biodiversity causing, in certain cases, even the death of animal and plant species. All this happens when the rate of degradation of the algae by microorganisms is greater than that of oxygen regeneration, which in summer is already present in low concentrations. The stage is eutrophic, when the lake is filled with sediment, while aquatic animal life will perish. It will then turn into dry land. The rate of eutrophication strikes a balance between the production of aquatic life and its destruction by bacterial decomposition. With large input of nutrients from human sources, bacterial decomposition cannot keep pace with productivity and sedimentation is accelerated whereby eutrophication is favoured. Lakes can be protected from eutrophication by providing measures for sewage treatment and preventing the sewage and agricultural runoff from entering the water bodies. Another method is to use aquatic plants as their high relative growth rates efficiently absorb nutrients from their surrounding media, thereby providing a simple and inexpensive solution for phosphorus-polluted water bodies. Experiments done by Shardendu et al. [34] proved that *Pistia stratiotes* L. accumulated the highest amount of tissue P (1 .06 ± 0 .22 mg/g dw) than other common wetland species like *Eichhornia*, *Phragmites* and *Typha*. They further found out that up to 91% phosphate was removed from the surrounding medium within 60 days at 50 mg/L supply.

Author details

D. Sayantan[1*] and Sumona Sanyal Das[2]

1 Department of Life Sciences, CHRIST (Deemed to be University), Bengaluru, Karnataka, India

2 Department of Zoology, Patna University, Patna, Bihar, India

*Address all correspondence to: sayantan.d@christuniversity.in

References

[1] Ahemad M, Zaidi A, Khan MS, Oves M. Biological importance of phosphorus and phosphate solubilizing microbes. In: Khan MS, Zaidi A, editors. Phosphate Solubilizing Microbes for Crop Improvement. NY, USA: Nova Science Publishers; 2011

[2] Helmenstine AM. Chemical Composition of the Earth's Crust—Elements. ThoughtCo. 2019. Available from: thoughtco.com/chemical-composition-of-earths-crust-elements-607576 [Accessed: 06 December 2019]

[3] Parkes GD, Mellor JW. Mellor's Modern Inorganic Chemistry. London, UK: Longman's Green and Co.; 1939

[4] Sommers AM. Phosphorus (Understanding the Elements of the Periodic Table). NY, USA: The Rosen Publishing Group; 2007. p. 25. ISBN: 1404219609

[5] Cai Y, Zhang G, Zhang Y-W. Phosphorene: Physical Properties, Synthesis and Fabrication. NY, USA: Jenny Stanford Publishing Pte. Ltd.; 2019. ISBN 978-981-4774-64-2 (Hardcover), 978-0-203-71061-6 (e Book). Available from: www.jennystanford.com

[6] Lumen. Introduction to Chemistry. Properties of Phosphorus. 2019. [Accessed: 06 December 2019]

[7] Los Alamos National Laboratory. Periodic Table of the Elements. 2019. Available from: https://depts.washington.edu/eooptic/linkfiles/The%20Elements.pdf [Accessed: 06 December 2019]

[8] Oliveira P. The Elements: Periodic Table Reference. Boppstrasse 64, Mainia, Germany: Pedia Press GmbH; 2011. Available from: http://pediapress.com

[9] Aviv U, Kornhaber R, Harats M, Haik J. The burning issue of white phosphorus: A case report and review of the literature. Disaster and Military Medicine. 2017;**3**:6. DOI: 10.1186/s40696-017-0034-y

[10] Phosphorus in Soil Crop System. The National Academics of Sciences Engineering Medicine. The National Academics Press. Washington DC, USA: Soil and Water Quality. An Agenda for Agriculture; 1993. p. 284. Available from: https://nap.edu/read/2132/chapter11#284

[11] Lappalainen HK et al. Pan-Eurasian Experiment (PEEX): Towards a holistic understanding of the feedbacks and interactions in the land-atmosphere-ocean-society continuum in the northern Eurasian region. Atmospheric Chemistry and Physics. 2016;**16**:14421-14461

[12] Zhang MK, Liu ZY, Wang H. Use of single extraction methods to predict bioavailability of heavy metals in polluted soils to rice. Communications in Soil Science and Plant Analysis. 2010;**41**(7):820-831

[13] Khan S, Cao Q, Zheng YM, Huang YZ, Zhu YG. Health risks of heavy metals in contaminated soils and food crops irrigated with wastewater in Beijing, China. Environmental Pollution. 2008;**152**(3):686-692

[14] Sood A, Uniyal PL, Prasanna R, Ahluwalia AS. Phytoremediation potential of aquatic macrophytes, Azolla. Ambio. 2012;**41**(2):122-137

[15] Ali H, Khan E, Ilahi I. Environmental chemistry and ecotoxicology of hazardous heavy metals: Environmental persistence, toxicity, and bioaccumulation. Journal of Chemistry. 2019; Article ID 6730305. DOI: 10.1155/2019/6730305

[16] Masindi V, Muedi KL. In: Saleh HE-DM, Aglan RF, editors. Environmental Contamination by Heavy Metals, Heavy Metals. IntechOpen; 2018. DOI: 10.5772/intechopen.76082. Available from: https://www.intechopen.com/books/heavy-metals/environmental-contamination-by-heavy-metals

[17] Gill SS, Tuteja N. Reactive oxygen species and antioxidant machinery in abiotic stress tolerance in crop plants. Plant Physiology and Biochemistry. 2010;**48**:909-930

[18] Maini G, Sharman AK, Knowles CJ, Sunderland G, Jackman SA. Electrokinetic remediation of metals and organics from historically contaminated soil. Journal of Chemical Technology and Biotechnology. 2000;**75**:657-664. DOI: 10.1002/1097-4660(200008)75:8<657::AID-JCTB263>3.0.CO;2-5

[19] Jones CJEP. Earth Reinforcement and Soil Structures. London: Thomas Telford Publishing; 1996. p. 379

[20] Lambert M, Leven BA, Green RM. New Methods of Cleaning Up Heavy Metal in Soils and Water. Available from: https://cfpub.epa.gov/ncer_abstracts/index.cfm/fuseaction/display.files/fileID/14295

[21] Comparing Costs of Remediation Technologies. National Research Council. Innovation in Groundwater and Soil Cleanup: From Concept to Commercialization. Washington DC: The National Academies Press; 1997. p. 10.17226/5781

[22] Peuke AD, Rennenberg H. Phytoremediation. EMBO Reports. 2005;**6**(6):497-501. DOI: 10.1038/sj.embor.7400445

[23] Sayantan D, Shardendu. Phosphate amendments moderate the arsenate accumulation and its subsequent oxidative and physiological toxicities in *Amaranthus viridis* L. Proceedings of the National Academy of Sciences, India, Section B: Biological Sciences. 2017;**87**:1343-1353. DOI: 10.1007/s40011-016-0711-5

[24] Sayantan D, Shardendu. Amendment in phosphorus levels moderate the chromium toxicity in *Raphanus sativus* L. as assayed by antioxidant enzymes activities. Ecotoxicology and Environmental Safety. 2013;**95**:161-170

[25] Sayantan D, Shardendu. Phosphorus amendment competitively prevents chromium uptake and mitigates its toxicity in *Spinacea oleracea* L. Indian Journal of Experimental Biology. 2015;**53**:395-405

[26] Elias M, Wellner A, Goldin-Azulay K, Chabrier E, et al. The molecular basis of phosphate discrimination in arsenate-rich environments. Nature. 2012;**491**:134-137. DOI: 10.1038/nature11517

[27] Zhao FJ, Ma JF, Meharg AA, McGrath SP. Arsenic uptake and metabolism in plants. The New Phytologist. 2009;**181**:777-794

[28] Qian H, Sun Z, Sun L, Jiang Y, Wei Y, Xie J, et al. Phosphorus availability changes chromium toxicity in the freshwater alga *Chlorella vulgaris*. Chemosphere. 2013;**93**(6):885-891. DOI: 10.1016/j.chemosphere.2013.05.035

[29] Central Pollution Control Board. Designated Best Use Water Quality Criteria. Forest and Climate Change, Government of India: Ministry of Environment; 2019. Available from: https://cpcb.nic.in/wqstandards/

[30] Available from: http://www.eniscuola.net/wp-content/uploads/2016/11/Pdf_eutrophication.pdf

[31] Carpenter SR. Submersed vegetation: An internal factor in lake

ecosystem succession. The American Naturalist. 1981;**118**:372-383

[32] De AK. Environmental Chemistry. 7th ed. Delhi: New Age International Publishers; 2010. p. 214

[33] Penelope RV, Charles RV. Water Resources and the Quality of Natural Waters. London: Jones and Bartbett publishers; 1992

[34] Shardendu S, Sayantan D, Sharma D, Irfan S. Luxury uptake and removal of phosphorus from water column by representative aquatic plants and its implication for wetland management. ISRN Soil Science. 2012;**2012**:Article ID 516947, 9 pages. DOI: 10.5402/2012/516947

Chapter 3

Bone Mineralisation

Pinki Dey

Abstract

The mineralisation term mentions the development of inorganic precipitation over an organic background. This process occurs in a life span of biological organism for the formation of bone, teeth, exoskeletons, egg shells, etc. So, basically bone mineralisation is defined as the process of deposition of minerals on the bone matrix for the development of bone. The human bone is made up of 60–70% minerals which include calcium phosphate in the form of hydroxyapatite followed by 20–40% organic matrix containing type I collagen fibres and less than 5% of water and lipids. During bone mineralisation process osteoblasts which are also known as bone forming cells, aids to the production of calcium phosphate crystals which are then aligned in the collagen based fibrous matrix. The bone mineralisation procedure also known as calcification is a lifelong activity of a human being.

Keywords: bone, mineralisation, calcium phosphates and collagen fibres

1. Introduction

Bone is a multifaceted system which behaves as mechanical shield for providing support and security. Bone also plays an important role in haemostasis. Recently, it came to observation that bone also aids in functioning of endocrine glands. To fulfil all these purposes, bone is a well architecture behaving as a functionally graded structure from millimetre to nanometre range. As a result of this gradation bone shows an unusual amalgamation of high stiffness and toughness which are frequently inversely associated. Broadly, bone is made up of organic and mineral part where the organic part is comprised of type I collagen whereas the mineral part contains the calcium deficient hydroxyapatite. The mineral part of the bone is interlinked with collagen in such an organised manner that it not only provides flexibility and ductility to the structure but also shows stiffness. This organisation of bone when observed at nanoscale range it behaves as a composite which acts as a shield to the brittle hydroxyapatite from damaging and also helps in carrying load by transferring forces around the bond hence thereby reducing the stress formation in the composite matrix. The fibres of the collagen intermingled with mineral part are arranged in very different manner when observed on microscopic range. This arrangement at micrometre level completely depends on the rate of bone formation or bone location on the substrate. This is due to the various functions played by the bone tissue, i.e., during fracture rapid bone formation development of bone during growth, unhurried bone formation in order to adjust in accordance to mechanical requirements so as to avoid any damage to its structure as well as maintaining its mechanical assets. Bone can be divided into two major categories depending on its mechanical and biological aspects i.e. cortical bone (compact/dense) and cancellous (porous/spongy bone namely trabecular) [1].

2. Bone composition

The mineral portion of the bone which basically comprises of hydroxyapatite contributes to the 65% of weight of bone. The remaining 20–30% of the bone weight composed of collagen of Type I. And, the last 10% of weight includes water molecules which are present collagen-mineral structure. Also, there are some free water molecules in the bone composite which gets redistributed during load bearing phase of bone. These free molecules pass through the vascular channels of bones and hence play a pivotal role in detection of signals by cells thus transferring information regarding load bearing environment. The interaction between bone mineral and unbound water molecules in a ratio of 1:1 helps in the process of bone mineralisation i.e. when the amount of water declines the mineralisation of bone starts to proceed and vice versa. The bone mineralisation is directly related to the stiffness of the bone as bone tends to grow stiffer since it contains higher amounts of mineral and lesser quantity of water. And as a result of this stiffened bone becomes more prone to brittleness and hence fracture easily. Approximately, 90% of organic part of bone is made up of collagen of type I category also collagen of type III and V are present in very minor quantities. And, the last 10% belong to non-collagenous proteins which help in the regulation of collagen development and management of fibre size, resistance to micro-crack, cellular adhesion and mineralisation. Around 85% of non-collagenous proteins are present in extracellular matrix and rest resides with the bone cells.

2.1 Collagen

The fibres of collagen are intertwined with the plate like structure of bone. Each and every collagen molecule exists as a triple helical structure formed from two chains of α 1 and one chain of α 2. Individual chain has around 1000 amino acids lengthwise and the centre part of the helix contains triplets of gly-X-Y in repeating sequence. The glycine molecules help in the formation of helical structure of collagen. Basically, all the amino acids are incorporated in collagen but the X and Y units are often composed of residue of hydroproline and proline. The maintenance of helical rigidity of the chain is due to the proline and hydroproline residues. The hydroxyl part of the hydroproline amino acid is important for interaction with water molecules via hydrogen bonding. The triple helix structure of collagen retained by the water molecules is affined toward hydroxyproline. At intercellular stage, peptides of non-helical region, i.e., N-propeptide as well as C-propeptides hold the chains at one place by cross- linking with sulphur. The propeptides present at the terminal point of triple helix are called procollagen molecule. During the exocytosis of molecules, the propeptide parts are broken down enzymatically resulting in non-helical parts at the molecular end N or C-terminal respectively. The enzymatically cleaved peptides result in the development of mature collagen molecule which has a pattern of non-helical N- and C-terminal peptides and helical nature of triple helix region. The microfibrils of collagen are semi-hexagonal system of five collagen molecules. The lateral and longitudinal combination of microfibrils leads to the formation of fibres approximately 10 μm in length and around 150 nm in diameter. The collagen fibres when observed under electron microscope showed a band pattern of around 67 nm. This band pattern is known as D-banding and it demonstrates the area between the neighbouring ends of collagen molecules as well as the overlapping between the lateral neighbouring molecules present at the end sections. In an osteoporotic bone the average diameter between the collagen fibrils and their spacing is very less as compared to a normal bone. The collagen fibrils are joined with the help of various types of cross-links which may have influential effects on the materialistic properties of the surrounding tissues and thereby affecting the mechanical traits of the whole bone system. The crosslinks bonds which connects the collagen fibrils are broadly categorised into enzymatic crosslinks and non-enzymatic crosslinks which forms a AGEs (advanced glycation end products).

2.2 Non collagenous proteins

There are various proteins of non-collagenous nature that helps in the regulation of formation and preservation of the extracellular matrix. Even though they comprise only of 2% of bone in weight but NCPs do a very important job during embryogenesis and formation as well as establishment of fibrils of collagen. It also regulates mineral of formation of bone and deliver channels for all signalling regarding the attachment of cells. The NCPs are divided into

1. *Proteoglycans*: They are basically consisting of various heparin sulphate, hyaluronan, small leucine rich proteoglycans called SLRPs. These are wide range of molecules where the main proteins are covalently bonded to the lateral chains of sulphated glycosaminoglycan. The proteoglycans which are present in the bone are smaller in size as compared to the other non-bone region proteoglycans. It manages the nucleation of apatite and the growth of apatite which in turn controls mineralisation.

2. *Glycoproteins*: It consists of fibronectin, thrombospondin (TSP1 and TSP2), vitronectin and lastly, alkaline phosphatase (ALP). In bones there are huge numbers of glycoproteins, out of which functions of few glycoproteins are not known. The activity of ALP helps in determining mineralisation factors because it undergoes hydrolysis with pyrophosphates which, as a result restricts the deposition of minerals by tagging itself with mineral crystals. If the pyrophosphates are neutralised, then this leads to normal growth of mineral crystals and as a result regulates the bone mineralisation. Since ALP is solely not produced by bone but also by kidney and liver any changes in the levels of ALP will not give any precise results regarding the mineralisation activity. Nonetheless, if the ALP is taken from the bone specific region, then the levels of ALP may determine the activity of bone mineralisation. In fact, lower levels of ALP or dysfunctionality of ALP leads to disorder recognised as hypophosphatemia which results in hypercalcemia causing death in children. TSP1 (Thrombospondin1) and TSP2 (Thrombospondin2) are found in mesenchymal stem cells and chondrocytes in the process of development of cartilage in the primary steps of bone development. TSP2 enhances the process of mineralisation as well as it escalates the osteoid process during mineralisation. Both vitronectin and fibronectin attaches to the cells where the vitronectin helps in attachment of cells which are found in plasma membrane of osteoclast and works with the osteopontin for binding osteoclasts to the matrix of mineral. The fibronectin plays an important role in cell proliferation during formation of bone.

3. *SIBLING* also known as small integrin binding ligand N linked glycoprotein which includes dentin matrix acidic phosphoprotein 1 (DMP1), osteopontin, sialoproteins and MEPE (matrix extracellular phosphoglycoprotein). In the preliminary stages of osteogenesis osteopontin is secreted. The ostoponin exists near the periodontal region of teeth as well as the cement line of bone. It restricts the crystal growth during mineralisation, also it attaches itself to the osteoclast in order to enhance the binding of osteoclast to the mineral surface of bone in the course of bone resorption. DEMP1 has an immense inclination

for hydroxyl apatite and N-telopeptide region of type I collagen, also is indicated by osteocytes and osteoblasts. It helps in local regulation of bone mineralisation. Although, it is quite an unknown fact whether DMP1 is involved in the differentiation process of osteoblasts to osteocytes. But in adequate DMP1 results in hypophosphatemia rickets due to increased level of FGF23. MEPE also belongs to SIBLING genre which functions to locally regulate Mineralisation. It is mainly found in osteocytes and odontoblasts, where it is supremely demonstrated in tissues which are still under Mineralisation such as modification of intramembrane and endochondral plaited bone of your fractured callus. *In vivo* results show that lack of MEPE gives increased bone mass as well as decreased in bone loss.

4. *Osteocalcin (Gla bone protein)*: The osteocalcin is indicated in osteoblasts as well as osteocytes. It helps in binding o calcium and deposition of mineral. As a result, it is considered as an indicator of bone formation. However sometimes It may also behave as a precursor in regulation of osteoclasts. It has been seen that mice in absentia of osteocalcin suffers from grave osteoporosis. Hence, it can be seen why osteocalcin is considered as one of the important marker for bone remodelling. It has been also observed that post-menopausal osteoporosis that its increased level increases bone remodelling rate causing acute imbalance between resorption and formation of bone.

5. *Osteonectin* are also known as SPARC, i.e., secreted proteins acidic rich in cysteine. It is present at mineral deposited location where it bonds with collagen, hydroxyapatite and vitronectin. It helps proliferation of freshly nucleated mineral crystals. Since its plays a vital part in osteoblasts growth, its non-existence causes osteopenia i.e. low bone density. It bandages itself to various growth factors such as FEF2, PDGF (platelet derived growth factor), VEGF (vascular endothelial growth factor) for the regulation of mineralisation.

3. Bone mineral

Bone mineral is made up of carbonate apatite which has poor crystalline structure. The apatite undergoes nucleation in the space between collagen fibrils end called hole zones and it spreads longitudinal over the collagen fibrils. In the beginning, mineral is settled in the form of amorphous calcium phosphate followed by calcium carbonate in good amount. The carbonate proportion reduces during bone tissue maturation, also mineral crystals form disc like structure growing laterally while aligning themselves parallel to the fibrils of collagen. The l-axis (long axis) of mineral disc oriented with the longitudinal line of bone. The typical size of mineral crystals has less than 10 nm thickness. Gradually, the mineral disc merge with other crystals to form a large polycrystalline structure, which becomes indeed larger than the thickness of fibrils. The mineral crystals grow more in size during bone ageing due to changes in ion contents of mineral composition. The age of the tissue is directly proportional to the size of the crystals. Although, it is tough to differentiate amid small crystals bearing numerous defects and large crystals having less defects, as both shows likewise crystalline behaviour. The soluble carbonate which adheres to surface of crystals can also be filled in by hydroxyl and phosphates groups of carbonate apatite. As a result, it helps in easy resorption of mineral. During the incidents of acid load, bicarbonate (-HCO_3) is being absorbed so as to maintain the pH of blood. The deficiency of -HCO_3 is balanced by the presence of carbonate and phosphate ions in bone mineral. When there is an abnormal production of acid, the bone mineral tank helps in the maintenance of acid–base balance, which also many times leads to loss in bone mass. There are certain cations such as Mg, Na, strontium in place of calcium ions and fluoride ions in place of hydroxyl ions in apatite matrix. Also, these kinds of substitutions can cause changes in mechanical properties of bone as well as in the behaviour of osteoclast and osteoblasts. Previously, for osteoporosis sodium fluoride used as an anabolic remedy. It was observed that sodium fluoride promoted pre-osteoblasts and osteoprogenitor cells hence stimulating uninterrupted formation of bone without the initialization of resorption. However, it has been seemed that carbonate apatite is less unaffected by resorption than fluoro-apatite. Also, replacement of fluoride ions into the mineral leads to escalated brittleness of bone thus causing the bone prone to fracture. The poor mechanical behaviour of bone is only due to the replacement of fluoride ions from bone as compared to the occurrence of substitutions of other ions from the bone mineral. The accumulation of bone mineralisation takes place in two consecutive phases. The first phase is the rapid nucleation of the primary mineral crystals. This phase is also known as primary Mineralisation. And, second phase relates to the slow proliferation and development of the primary crystals up to a size of 40 × 3 × 7.5 nm. In the course of primary mineralisation crystals are very quickly accumulated in the collagen network thus attaining 65–70% of total mineralisation within 3 weeks approximately. During the second step, the mineral is deposited at a steady rate but in a more efficient way, till the mineral attains the required bodily limit, which may vary from months to year [1–19].

4. Bone cells

Bone is a structurally and metabolically very complicated organ which is a composite of mineral, collagen material and bone cells [20]. The bone cells basically include osteoblasts, osteoclasts and osteocytes, which are found in mesenchymal stem cells known to accumulate osteiod before the mineralisation process takes off, thereby helping in bone formation [21, 22]. Osteoblasts are known for regulating mineralisation and in the formation of extracellular matrix. They are originated from bone and are in cuboidal form especially found at bone surface and carries out the function of resorption [21, 23]. The quantity and function of osteoclasts are dependent on many factors such as proliferation, differentiation, rate of resorption by already developed osteoclasts and lastly cell lineage allocation [24]. They are the derivative of multinucleated polarised cells which are migratory in nature with good source of lysozyme enzymes [25]. They consist of mitochondria of pleomorphic type, vacuoles and lysosomes [26]. The formation and resorption of bone is the joint activity of osteoclasts and osteoblasts. And the factors which are involved during this process are prostaglandin E2 (PGE2), transforming growth factor beta 1 (TGF-β1), fibroblast growth factor, parathyroid hormone (PTH), osteoprotegerin ligand OPGL also known as RANKL (receptor activator of nuclear factor kappa B [(NF-κB) ligand]) and TRANCE i.e. TNF related activation induced cytokine. Resorption can also effect the biomechanical activities of bone for instance, formation of strong bone from a weak one [24]. Majority of bone cells are in the nature of osteocytes, thus comprising of 90–95% in the skeleton of an adult. The mature osteoblasts in the bone matrix are recognised as osteocytes. And they help in responding to the mechanical strain thus generating signals which can further coordinate the bone resorption and its formation [27]. Osteocytes that are present in mature bone are joined together with long extensions of cytoplasm that form small capillary like structure called lacunae or canalcali for the transfer of nutrients and wastes. Osteocytes are spread across the mineral matrix and connect to the surface of bone and bone marrow via dendrites which involve osteoclast precursors for the stimulation of bone resorption and regulation of differentiation of mesenchymal stem cells [28–32]. The roles which are played by osteocytes and lacunae/ canalcali involves restriction of fatigue cracks, exchange of mineral, hormonal stimulation for detection of stress or strain, mending of microdamage, modelling or remodelling of bone under mechanical criteria, osteocytic osteolysis, regulation of osteoclastic cutting cone during exchange of mineral and reformed remodelling act after the resorption [27].

During the bone mineralisation, the crystals of mineral are accumulated in a systematic manner over the extracellular matrix, where the cells surrounding the mineral matrix prepares a pattern for mineral accumulation thus commencing the location for mineralisation and fixing the final dimensions of mineral crystals. Although, various studies have been conducted throughout the world for determining the mechanism for formation of mineral crystals in every organism, but the exact explanation related to this mechanism remains unclear [33–35]. In accordance to the conventional theories about biomineralization, NCPs were actively engaged in the process of matrix mineralisation. In 1994, Hunter and Goldberg postulated that the effects of mineralisation for BSPs were completely connected to the glutamate and aspartate-rich sequences of peptides [36]. Later around 1997, Stubbs et al. carried out studies to consider the involvement of other groups such as sulphate, phosphate sialic acidic groups in the process of mineralisation [37]. According to earlier reports, nucleation of mineral takes place in their principal ionic solution which is supersaturated in nature. During nucleation in solid phase, a critical size of crystal is required in order to initiate the nucleation. This mechanism is known as stochastic solute clustering [38, 39].

At present, two important models have been considered for bone mineralisation process. The first model which involves mineralisation with the help of collagen template and the second model include the matrix vesicles for mineralisation purpose. It has been widely accepted that mineral formation is a systematic procedure which can never take place in the absence of matrix. Basically, the matrix gives an ordered pattern of deposition of mineral, thus directly participating in the mineralisation by behaving as a nuclear. It has been also seen that different mineralised tissues have different matrices [40]. Alternatively, matrix vesicles are the particles derived from extracellular matrix having a diameter of 100 nm, precisely positioned inside the bone matrix and the matrix of cartilage and peridentin. They provide the initial location for the calcification of all skeletal tissues. They are generally formed from a polarised bud which gets discharged from surface of the chondrocytes, osteoblasts and odontoblasts [41, 42]. The matrix vesicles are considered for preliminary location for mineral build-ups in bone tissue [43, 44]. In course of mineralisation involving cells, the formation of primary hydroxyapatite crystals takes place inside the vesicle membrane matrix [45].

a. *Collagen-moderated mineralisation*: In this type of mineralisation, the template for accumulation of mineral is provided by the collagen present in the bone tissue. And, these very collagen fibrils decide the sizes of crystals that can attained for the process of mineralisation. On the other hand, mineralisation does not take place in a deficiency of NCPs because they behave as molecules that generate signals all through the course of mineralisation [40, 46]. It was observed that BSPs role as a crystal nucleator, affected the osteocalcin recognition and remodelling of mineralised surfaces [36, 37, 47–51]. However, osteopontin and osteonectin helped in regulation of crystal formation on the basis of size, type and growth [52, 53]. During the growth phase, the crystals nucleates from an amorphous phase were the intervallic pattern of 67 nm cross-striated collagen fibres [54, 55] carryover the nucleation in the 40 nm long gap sparsely dense zone. The aforementioned process was recommended to be guided by heavily acidic NCPs [56–59]. The basic principal of collagen based template mineralisation focuses mainly on the job of collagen fibrils during bone Mineralisation. When the mineralising fibre were observed under cryo-transmission electron microscope, it was found that polyaspartic acid which has soluble behaviour plays an integral part in collagen mineralisation [56, 60]. The calcium triphosphate ions complex is formed from the prenucleated clusters of acidic polypeptide [38]. These clusters are negatively charged; as a result, they get attracted towards the positive part on collagen [56]. Consequently, these ionic complex gets fused inside the collagen fibrils and then transform into a solid amorphous mass which further grows into an ordered apatite crystal complex regulated by the arrangement of collagen fibrils [40].

b. *Matrix-vesicle moderated mineralisation*: In this type of mechanism, the preliminary phase starts off at the mineral visceral where the Ca^{2+} ions and the inorganic phosphate (Pi) are formed *in vivo* [41, 61, 62]. The annexins and phosphatidylserine are the calcium binding molecules which tag them with BSPs so as to invite and regulate the deposition of calcium and phosphate ions previous to the creation of crystals of insoluble hydroxyapatite [62, 63]. During, this phase the pH of the intravesicules rises above due to the activity of carbonic anhydrase found in mineral vesicles [64], causing the stabilisation of primary mineral crystals [42]. In the second phase of matrix moderated mineralisation, the breakdown of mineral vesicles membranes takes place where the already formed hydroxyapatite are exposed to extracellular fluid [61]. The extracellular fluid comprises of matrix vesicles with homeostatically regulated levels of PO_4^{3-} and Ca2+ in order to help in proliferation of new hydroxyapatite crystals onto the already formed hydroxyapatite crystals. The perforations in the matrix vesicle membrane are carried out by proteases [65] and phospholipases [66]. The metalloproteinases of matrix vesicles which have the capability of degenerating mineral deficient proteoglycans helps in transferring of mineral towards itself [67]. In the recent studies, it has been observed that collagen type II and X bind to the outside surfaces of matrix vesicles thus acting as a channel for the transfer of crystals into the extravesicular matrix [68]. Three promising functions of matrix vesicles have been identified during the course of mineralisation. The first function involves the control of ion concentration by the matrix vesicles inside the matrix so as to start off the mineralisation around the collagen fibrils. Also, it controls the compositions of ions necessary for the formation of intravesicular apatite crystals thus starting the process of mineralisation with the transfer of ions to the collagen. And lastly, when the mineral vesicles interact with the collagen, the deposition of mineral onto the surface of fibrils is carried out [40].

5. Pathological mineralisation

Mineralisation is categorised into physiological or pathological types depending on the type of bone tissues i.e. hard bone tissue or soft bone tissue. Physiological mineralisation is required for the development of skeletal tissues in order to carry out daily functions of a normal human life. The second category of mineralisation is pathological mineralisation also known as ectopic which involves the mineralisation of soft bone tissues such as cartilages (articular cartilage) and tissues surrounding cardiac vessels that causes diseases and death. Recently, it has been found that the reasons and aspects which cause physiological mineralisation can be similar to that of pathological. Lately, it has been reported both the mineralisation are instigated by the matrix vesicles where the particles which resides inside the membrane are released from the plasma membrane of mineralisation cells. The activators and regulators which cause pathological mineralisation are the same activators and regulators which will cause pathological mineralisation. It has been also reported that apoptosis causes physiological mineralisation and it has been seen that if physiological mineralisation happens to happen take place after injury of tissue, can prompt pathological mineralisation around the damaged or the injured tissue [69]. The mineralisation process is important for imparting mechanical properties in the bone [70]. As a result, when not regulated properly can lead to the inadequate or extreme mineralisation. Therefore, bone tissue quality gets jeopardised and becoming the cause for many bone related diseases. Osteomalacia is one such condition where the disease is caused in adults due to deficiency of bone mineral or excessive bone resorption. Rickets is the osteomalacia in children. As already explained, in a fit and mature bone, osteoclasts eliminate bone whereas osteoblasts accumulate osteoid and thus carrying out the mineralisation. During osteomalacia, calcification rate is decreased while the bone surface is being increased due to the building up of non-mineralised osteoid. The common symptoms of osteomalacia involves brittle bone, weakened muscles along with severe body ache [71–73]. Another known disease caused by pathological mineralisation is fibrous osteodystropy. In various findings, it has been seen that the flexibility and deformation of bone depends because of constant and extreme contact with the PTH, thus hampering with the bone load bearing capacity [74–78]. In osteocalcia, bone tissue is destroyed and resorped osteoclasts where the remodelling area, previously taken up by calcified bone is then occupied by fibrous connective tissue. As the disease progresses, non-mineralised formed bone takes up the place of cortical bone. And also, mineralised osteoid in the remodelling space was previously filled up by osseous tissue [79]. Paget's disease is a very commonly known disorder of bone in adults which is chronic in nature. This is also known as osteitis deformans [80, 81]. This is mainly found in middle aged men as compared to women [80, 82, 83]. In this disease, the resorption mechanism of bone gets speed up which results in formation of thick unarranged bone mineral matrix. As a result, producing weakened bone structure, painful fragile bone, joints arthritis inside the targeted bone. Sometimes, Paget's disease transform into a preliminary cancer of bone identified as Paget's sarcoma. The large multinucleated osteoclasts initiate the pathological Mineralisation of bones with high resorption causing Paget's disease [84]. This hastened resorption results in an unorganised deposition of mineral by osteoblasts during remodelling [85]. Thus causing irregularities in cortical thickness, coarsening of trabecular and vascularization of fibrous tissues, which then produces fatigue during high stress condition [81, 86, 87]. Another very well-known bone disease is osteoporosis, where the affected bone has depleted mass with structural degradation as well as amplified porosity of bone tissue [88]. Usually, in osteoporosis all bones are affected as compared to the Paget's disease where only a part of bone is targeted. When the mineral content of the bone goes below the critical value, the bone becomes more brittle in nature thus the load bearing capacity along with other mechanical properties of bone gets deteriorated [89]. The BMD also known as bone mineral density is directly proportional to the mechanical strength of bone. The patients with ongoing history of osteoporosis has reduced BMD, hence are more prone to fracture [90–93]. There are basically three main reasons behind the cause of osteoporosis. The first reason postulates that the commencing of the osteoporosis may be due to the underdevelopment of bone during the growth period of individuals. The second reason focuses on the bone development due to heightened resorption process.

And lastly, osteoporosis may happen due to the lack of new bone growth during the course of remodelling [94]. Earlier, it was believed that osteoporosis is an outbreak of ageing process but with the recent studies it has been observed that it may be caused because of malnutrition, alterations in biomechanical loading, production of excessive hormones and prolonged history of acidosis [95].

6. Mineralisation of synthetic biomaterials

For many years, biomaterials are being experimented in such a way so as to choose them as a replacement for damaged or diseased tissues. As the exclusive qualities of bone tissues are solely connected to the bone mineralisation, thereby interpreting and regulating the mineralisation process of artificial bone replacement is very crucial [20].

6.1 Inorganic materials

Since the mineral part of the bone is inorganic in nature, many biomaterials such as calcium phosphates, hydroxyapatite and bioglass are being used as bone alternatives [96]. It was in 1969, it was observed that bioglass forms bond with the bone thus restricting the development of fibrous tissues surrounding the bone [97]. There are various other ceramic biomaterials available such as hydroxyapatite [98], β-tricalcium phosphate, glass–ceramic [99–101], which are after sintering have displayed the bone bonding ability, hence paving the way for themselves clinically in reconstructive and regenerative medicine fields [102].

6.2 Calcium phosphates

The first in the list of very commonly used bioceramics are the calcium phosphate based ceramics, broadly used in the field of dentistry and orthopaedics. They are mainly used for purpose of coatings onto the top surface of metal implants such as titanium, etc. The study of mineralisation can be performed in some fluids which imitates the ionic composition of blood plasma. One such kind of fluid is simulated body fluid suggested by Kokubo et al. in 1991. This fluid mimics the ionic concentrations of human blood plasma almost exactly. This fluid is used for the purpose of studying mineralisation behaviour of biomaterials in vitro. The SBF initiates the formation of apatite (bone-like) over the synthetic biomaterial surfaces, thus giving the idea of mineralisation of the biomaterial in vivo [103]. Kokubo et al. also suggested that in vivo bone reaction of an artificial biomaterial can be anticipated by the formation of apatite over its surface in SBF, but this theory now has been challenged [104]. In 2010, Bertazzo et al. suggested that calcium phosphates exhibited the osteoconductive nature of bone guided by bone tissues beside the biomaterial surface at the site of implantation orthotopically. Also, there are other calcium phosphate ceramics present which shows osteoinductive traits i.e. ability to form bone at implantation site ectopically. Although, the exact mechanism behind the osteoinductive nature of certain calcium phosphates depend on various factors such as composition, structure and conformation of the calcium phosphate biomaterial [105].

6.3 Bioactive glasses

In 1969, Hench observed that certain silicon-based glasses formed bond with the bone [106] and named it as Bioactive® glass also known as 45S5 consisting of

45 wt% of SiO_2, 24.5 wt% of CaO, 24.5 wt% of Na_2O and 6 wt% of P_2O_5.it can be synthesised by two methods which are sol-gel and melt-quench. Because of its awesome response towards the bone formation in vivo, it has been considered for more clinical applications in the arena of orthopaedics. To comprehend the behaviour of bioglasses in vivo, they have been submerged in SBF to analyse the physio-chemical route of Mineralisation onto the material surface. It has been found that a carbonated apatite layer is formed which is almost like bone mineral. The mineralisation behaviour of calcium phosphate ceramics and bioglasses follow the same mechanism for the apatite formation, thus establishing a bond (chemically) between the host bone and biomaterial [107].

6.4 Organic materials

Moreover, along with inorganic biomaterials there are other organic biomaterial in the form of biopolymers are available which also shows excellent properties to be considered for tissue regeneration. Such kind of polymers is PLA (polylactic acid), PGA (polyglycolic acid), collagen, hyaluronic acid and many more. The biopolymers do not help in the formation of bone as that of bioceramics but they act as a supporting matrix for the treatment of damaged bone. The biopolymers are broadly divided into two categories I.e. Hydrated and non-hydrated polymers depending on the water retention ability [20].

6.5 Hydrated biopolymers

Hydrogels are considered under this label as the water is taken up by the polymeric network making them swollen in shape. Since they can take up water very quickly and can retain it, they are considered mainly for the application of cell culture, drug delivery and tissue engineering. On the other hand, they cannot be considered for bone regeneration purpose as they cannot provide any mechanical stability to the affected site [108, 109]. Also, the water content of hydrogels gets so high that it becomes almost impossible to sterilise them [110]. Furthermore, they fail to develop any with the surrounding tissues. The researchers around the world are trying to initiate mineralisation in the hydrogels by incorporating bioceramic particles such as calcium phosphates, hydroxyapatite or bioglass. Also, introduction of enzymes which catalyses the mineralisation activity into the hydrogels or injecting certain artificial analogues to the matrix vesicles in order to trigger biominerlisation. And, lastly the polymeric hydrogels are charged with negative ions or groups in order to invite positively charged calcium ions and thus stimulating mineralisation in inert hydrogels [110].

6.6 Non-hydrated biopolymers

This category includes many biopolymers such as PLA, PGA, collagen, chitosan, etc. to be considered broadly in the field of tissue engineering. The scaffolds which are made up of non-hydrated polymers can be developed by various techniques such as 3D printing [111], porogen leaching [112], fibre meshing, microsphere sintering [111], phase separation gas foaming [113], and supercritical fluid processing [114]. The scaffold so developed by aforementioned processing techniques differ in surface properties and porosities [115]. Mineralisation of these polymers can be attained by various ways such as incubating in SBF, modifying surface with anionic groups so as to attract calcium ions over the surface of biomaterial. In non-hydrated polymers the assessment of mineralisation behaviour can be studied using SBF similar to bioceramic scaffolds. The nature of SBF taken for mineralisation activity impacts the ionic composition and the configuration of mineral phase for example more concentrated form of SBF speeds up the bone mineralisation. The negatively charged proteins play integral role in controlling the deposition of mineral phase onto the natural bone, thus affecting the mineralisation in SBF. Hence, the surface of biopolymer is functionalized with negatively charged proteins or bone related proteins so as to trigger biominerlisation [116–120].

7. Conclusions

Bone is a very complex organ which has hierarchical organised tissue that helps in giving protection and support mechanically to all the organs including brain because of its mineralised behaviour. The mineralisation of bone is mediated by either collagen or matrix vesicles. Both the pathways are interconnected; but, their interconnectivity is still not extensively studied. By studying the mineralisation pattern of bone, several bone related diseases caused due to pathological mineralisation or other reasons lacking in the mineralisation can be cured with the help of development of advanced biomaterials which may showcase equivalent levels of biological functioning when compared to the natural bone.

Author details

Pinki Dey
Department of Ceramic Engineering, National Institute of Technology, Rourkela, Odisha, India

*Address all correspondence to: pinkideyvs@gmail.com

References

[1] Burr DB. Bone morphology and organization. In: Basic and Applied Bone Biology. 2nd ed. Academic Press, Elsevier; 2019

[2] Bonnucci E, Motta PM. Ultrastructure of skeletal tissues. In: Bone and Cartilage in Health and Disease. Boston: Kluwer Academic Publishers; 1990

[3] Brookes M, Revell WJ. Blood Supply of Bone: Scientific Aspects. London: Springer-Verlag; 1998

[4] Burr DB, Allen MR. Calcified tissue international, special issue: Bone material properties and and skeletal fragility. Calcified Tissue International. 2015;**97**:199-241

[5] Castañeda-Corral G, Jimenez-Andrade JM, Blook AP, Taylor RN, Mantyh WG, Kaczmarska MJ. The majority of myelinated and unmyelinated sensory nerve fibers that innervate bone express the tropomyosin receptor kinase A. Neuroscience. 2011;**178**:196-207

[6] Dempster D, Felsednberg D, van der Geest S. The Bone Quality Book. Amsterdam: Elsevier; 2006

[7] Enlow DH, Brown SO. A comparative histological study of fossil and recent bone tissues. Part III. Mammalian bone tissues. Texas Journal of Science. 1957;**10**:187-230

[8] Fonseca H, Moreira-Gonçalves D, Appell Coriolano HJ, Duarte JA. Bone quality: The determinants of bone strength and fragility. Sports Medicine. 2014;**44**:37-53

[9] Foote JS. A contribution to the comparative histology of the femur. Smithsonian Contrib. Knowl. 1916;**35**:1-242

[10] Fuchs RK, Allen MR, Ruppel ME, Diab T, Phipps RJ, Miller LM. In situ examination of the time-course for secondary mineralization of Haversian bone using synchrotron Fourier transform infrared microspectroscopy. Matrix Biology. 2008;**27**:34-41

[11] Fukumoto TJ. Bone as an endocrine organ. Trends Endocrinol. Metab. 2009;**20**:230-236

[12] Gurkan UA, Akkus O. The mechanical environment of bone marrow: A review. Annals of Biomedical Engineering. 2008;**36**:1978-1991

[13] Jee WSS, Weiss L. The skeletal tissues. In: Weiss L, editor. Histology: Cell and Tissue Biology. New York: Elsevier Biomedical; 1983

[14] Kaplan FS, Hayes WC, Keaveny TM, et al. Form and function of bone. In: Simon SR, editor. Orthopaedic Basic Science. Chicago: American Academy of Orthopaedic Surgeons; 1994

[15] Karsenty G, MacDougald O, Rosen CJ. Interactions between bone, adipose tissue and metabolism. Bone. 2012;**50**(Special Issue):429-579

[16] Martin RB, Burr DB, Sharkey NA, Fyhrie DP. Skeletal Tissue Mechanics. 2nd ed. New York: Springer-Verlag; 2015

[17] Reznikov N, Bilton M, Lari L, Stevens MM, Kröger R. Fractual-like hierarchical organization of bone begins at the nanoscale. Science. 2018;**360**:507-517

[18] Ruppel ME, Miller LM, Burr DB. The effect of the micro¬scopic and nanoscale structure on bone fragility. Osteoporos. Int. 2008;**19**:1251-1265

[19] Sivaraj KK, Adams RH. Blood vessel formation and function in bone. Development. 2016;**143**:2706-2715

[20] An J, Leeuwenburgh S, Wolke J, Jansen J. Mineralization processes in hard tissue: Bone. In: Biomineralization and Biomaterials. Elsevier; 2016. pp. 129-146

[21] Sommerfeldt D, Rubin C. Biology of bone and how it orchestrates the form and function of the skeleton. European Spine Journal. 2001;**10**(2):S86-S95

[22] Compton JT, Lee FY. A review of osteocyte function and the emerging importance of sclerostin. The Journal of Bone and Joint Surgery. American Volume. 2014;**96**(19):1659-1668

[23] Ferrari SL et al. A role for N-cadherin in the development of the differentiated osteoblastic phenotype. Journal of Bone and Mineral Research. 2000;**15**(2):198-208

[24] Harada S-I, Rodan GA. Control of osteoblast function and regulation of bone mass. Nature. 2003;**423**(6937):349-355

[25] Teitelbaum SL. Bone resorption by osteoclasts. Science. 2000;**289**(5484):1504-1508

[26] Walker DG. Enzymatic and electron microscopic analysis of isolated osteoclasts. Calcified Tissue Research. 1972;**9**(1):296-309

[27] Lanyon L. Osteocytes, strain detection, bone modeling and remodeling. Calcified Tissue International. 1993;**53**(1):S102-S107

[28] Bonewald LF. Osteocytes as dynamic multifunctional cells. Annals of the New York Academy of Sciences. 2007;**1116**(1):281-290

[29] Kamioka H, Honjo T, Takano-Yamamoto T. A three-dimensional distribution of osteocyte processes revealed by the combination of confocal laser scanning microscopy and differential interference contrast microscopy. Bone. 2001;**28**(2):145-149

[30] Baylink D et al. Vitamin D-enhanced osteocytic and osteoclastic bone resorption. The American Journal of Physiology. 1973;**224**(6):1345-1357

[31] Heino TJ, Hentunen TA, Väänänen HK. Conditioned medium from osteocytes stimulates the proliferation of bone marrow mesenchymal stem cells and their differentiation into osteoblasts. Experimental Cell Research. 2004;**294**(2):458-468

[32] Zhao S et al. MLO-Y4 osteocyte-like cells support osteoclast formation and activation. Journal of Bone and Mineral Research. 2002a;**17**(11):2068-2079

[33] Beniash E. Biominerals—Hierarchical nanocomposites: The example of bone. Wiley Interdisciplinary Reviews - Nanomedicine and Nanobiotechnology. 2011;**3**(1):47-69

[34] Boskey AL. Matrix proteins and mineralization: An overview. Connective Tissue Research. 1996;**35**(1-4):357-363

[35] Boskey AL. Biomineralization: Conflicts, challenges, and opportunities. Journal of Cellular Biochemistry. 1998;**72**(S30-31):83-91

[36] Hunter GK, Goldberg HA. Modulation of crystal formation by bone phosphoproteins: Role of glutamic acid-rich sequences in the nucleation of hydroxyapatite by bone sialoprotein. The Biochemical Journal. 1994;**302**:175-179

[37] Stubbs JT et al. Characterization of native and recombinant bone sialoprotein: Delineation of the mineral-binding and cell adhesion domains and structural analysis of the RGD domain. Journal of Bone and Mineral Research. 1997;**12**(8):1210-1222

[38] Gebauer D, Völkel A, Cölfen H. Stable prenucleation calcium carbonate clusters. Science. 2008;**322**(5909):1819-1822

[39] Pouget EM et al. The initial stages of template-controlled $CaCO_3$ formation revealed by cryo-TEM. Science. 2009;**323**(5920):1455-1458

[40] Chai YC et al. Current views on calcium phosphate osteogenicity and the translation into effective bone regeneration strategies. Acta Biomaterialia. 2012;**8**(11):3876-3887

[41] Anderson HC. Molecular biology of matrix vesicles. Clinical Orthopaedics and Related Research. 1995;**314**:266-280

[42] Anderson HC. Matrix vesicles and calcification. Current Rheumatology Reports. 2003;**5**(3):222-226

[43] Landis W. Chemistry and Biology of Mineralized Tissues. Toronto: University of Toronto Press; 2005

[44] Aparicio S et al. Optimal methods for processing mineralized tissues for Fourier transform infrared microspectroscopy. Calcified Tissue International. 2002;**70**(5):422-429

[45] Boskey AL. Mineralization of bones and teeth. Elements. 2007;**3**(6):385-391

[46] Palmer LC et al. Biomimetic systems for hydroxyapatite mineralization inspired by bone and enamel. Chemical Reviews. 2008;**108**(11):4754-4783

[47] He G et al. Phosphorylation of phosphophoryn is crucial for its function as a mediator of biomineralization. The Journal of Biological Chemistry. 2005;**280**(39):33109-33114

[48] Monfoulet L et al. Bone sialoprotein, but not osteopontin, deficiency impairs the mineralization of regenerating bone during cortical defect healing. Bone. 2010;**46**(2):447-452

[49] Farbod K et al. Interactions between inorganic and organic phases in bone tissue as a source of inspiration for design of novel nanocomposites. Tissue Engineering. Part B, Reviews. 2013;**20**(2):173-188

[50] Meyer U et al. Decreased expression of osteocalcin and osteonectin in relation to high strains and decreased mineralization in mandibular distraction osteogenesis. The Journal of Cranio-Maxillofacial Surgery. 1999;**27**(4):222-227

[51] Hoang QQ et al. Bone recognition mechanism of porcine osteocalcin from crystal structure. Nature. 2003;**425**(6961):977-980

[52] Roach H. Why does bone matrix contain non-collagenous proteins? The possible roles of osteocalcin, osteonectin, osteopontin and bone sialoprotein in bone mineralisation and resorption. Cell Biology International. 1994;**18**:617-628

[53] Rosenthal AK et al. Osteopontin promotes pathologic mineralization in articular cartilage. Matrix Biology. 2007;**26**(2):96-105

[54] Miller BF et al. Coordinated collagen and muscle protein synthesis in human patella tendon and quadriceps muscle after exercise. The Journal of Physiology. 2005;**567**(3):1021-1033

[55] Orgel JP et al. Microfibrillar structure of type I collagen in situ. Proceedings of the National Academy of Sciences of the United States of America. 2006;**103**(24):9001-9005

[56] Nudelman F et al. The role of collagen in bone apatite formation in the presence of hydroxyapatite nucleation inhibitors. Nature Materials. 2010;**9**(12):1004-1009

[57] Glimcher M, Muir H. Recent studies of the mineral phase in bone and its possible linkage to the organic matrix by protein-bound phosphate bonds. Philosophical Transactions of the Royal Society B. 1984;**304**(1121):479-508

[58] Landis W et al. Mineral and organic matrix interaction in normally calcifying tendon visualized in three dimensions by high-voltage electron microscopic tomography and graphic image reconstruction. Journal of Structural Biology. 1993;**110**(1):39-54

[59] Traub W, Arad T, Weiner S. Origin of mineral crystal growth in collagen fibrils. Matrix. 1992;**12**(4):251-255

[60] Cölfen H. Biomineralization: A crystal-clear view. Nature Materials. 2010;**9**(12):960-961

[61] Stewart T. The presence of delayed hypersensitivity reactions in patients toward cellular extracts of their malignant tumors. 1. The role of tissue antigen, nonspecific reactions of nuclear material, and bacterial antigen as a cause for this phenomenon. Cancer. 1969;**23**(6):1368-1379

[62] Wu L et al. Induction of mineral deposition by primary cultures of chicken growth plate chondrocytes in ascorbate-containing media. Evidence of an association between matrix vesicles and collagen. The Journal of Biological Chemistry. 1989;**264**(35):21346-21355

[63] Hunter GK, Goldberg HA. Nucleation of hydroxyapatite by bone sialoprotein. Proceedings of the National Academy of Sciences of the United States of America. 1993;**90**(18):8562-8565

[64] Stechschulte Jr, DJ, et al. Presence and specific concentration of carbonic anhydrase II in matrix vesicles. Bone and Mineral. 1992;**17**(2):187-191

[65] Hirschman A et al. Neutral peptidase activities in matrix vesicles from bovine fetal alveolar bone and dog osteosarcoma. Calcified Tissue International. 1983;**35**(1):791-797

[66] Wuthier RE. The role of phospholipids in biological calcification: Distribution of phospholipase activity in calcifying epiphyseal cartilage. Clinical Orthopaedics and Related Research. 1973;**90**:191-200

[67] Dean DD et al. Matrix vesicles are enriched in metalloproteinases that degrade proteoglycans. Calcified Tissue International. 1992;**50**(4):342-349

[68] Wu L et al. Collagen-binding proteins in collagenase-released matrix vesicles from cartilage. Interaction between matrix vesicle proteins and different types of collagen. The Journal of Biological Chemistry. 1991;**266**(2):1195-1203

[69] Kirsch T. Determinants of pathological mineralization. Current Opinion in Rheumatology. 2006;**18**(2):174-180

[70] Yeni Y, Brown C, Norman T. Influence of bone composition and apparent density on fracture toughness of the human femur and tibia. Bone. 1998;**22**(1):79-84

[71] Parfitt A. Osteomalacia and related disorders. In: Metabolic Bone Disease and Clinically Related Disorders. 2nd ed. Philadelphia, PA: WB Saunders; 1990. pp. 329-396

[72] Feng JQ et al. Loss of DMP1 causes rickets and osteomalacia and identifies a role for osteocytes in mineral metabolism. Nature Genetics. 2006;**38**(11):1310-1315

[73] Aaron J et al. Frequency of osteomalacia and osteoporosis in fractures of the proximal femur. Lancet. 1974;**303**(7851):229-233

[74] Lynch MJ et al. Fibrous osteodystrophy in dromedary camels

(Camelus dromedarius). Journal of Zoo and Wildlife Medicine. 1999;**30**:577-583

[75] Fincham J et al. Mseleni joint disease. Part I. An animal model? South African Medical Journal. 1985;**67**(2):51-57

[76] Long GG et al. Fibrous osteodystrophy in an opossum. Journal of Wildlife Diseases. 1975;**11**(2):221-223

[77] Flom JO, Brown RJ, Jones RE. Fibrous osteodystrophy in a wild dolphin. Journal of the American Veterinary Medical Association. 1978;**173**(9):1124-1126

[78] Jaffe H, Bodansky A, Blair JE. Fibrous osteodystrophy (osteitis fibrosa) in experimental hyperparathyroidism of Guinea-pigs. Archives of Pathology. 1931;**11**:207

[79] Fetter A, Siemering G, Riser W. Osteoporosis and osteopetrosis. In: Newton CD, Nunamaker DM, editors. Textbook of Small Animal Orthopaedics. Philadelphia, PA: Lippincott; 1985. pp. 627-629

[80] Whyte MP. Paget's disease of bone. The New England Journal of Medicine. 2006;**355**(6):593-600

[81] Siris ES. Paget's disease of bone. Journal of Bone and Mineral Research. 1998;**13**(7):1061-1065

[82] Kanis JA. Pathophysiology and Treatment of Paget's Disease of Bone. Oxford: Taylor & Francis; 1998

[83] Altman RD et al. Influence of disodium etidronate on clinical and laboratory manifestations of Paget's disease of bone (osteitis deformans). The New England Journal of Medicine. 1973;**289**(26):1379-1384

[84] Rebel A et al. Osteoclast ultrastructure in Paget's disease. Calcified Tissue Research. 1976;**20**(1):187-199

[85] Meunier PJ et al. Bone histomorphometry in Paget's disease: Quantitative and dynamic analysis of pagetic and nonpagetic bone tissue. Arthritis and Rheumatism. 1980;**23**(10):1095-1103

[86] Merkow R, Lane J. Paget's disease of bone. The Orthopedic Clinics of North America. 1990;**21**(1):171-189

[87] Ooi C, Fraser W. Paget's disease of bone. Postgraduate Medical Journal. 1997;**73**(856):69-74

[88] Kanis J et al. The use of clinical risk factors enhances the performance of BMD in the prediction of hip and osteoporotic fractures in men and women. Osteoporosis International. 2007;**18**(8):1033-1046

[89] Gao H et al. Materials become insensitive to flaws at nanoscale: Lessons from nature. Proceedings of the National Academy of Sciences of the United States of America. 2003;**100**(10):5597-5600

[90] Faibish D, Ott SM, Boskey AL. Mineral changes in osteoporosis: A review. Clinical Orthopaedics and Related Research. 2006;**443**:28

[91] Cefalu CA. Is bone mineral density predictive of fracture risk reduction? Current Medical Research and Opinion. 2004;**20**(3):341-349

[92] McCalden RW, McGeough JA. Age-related changes in the compressive strength of cancellous bone. The relative importance of changes in density and trabecular architecture. The Journal of Bone and Joint Surgery. American Volume. 1997;**79**(3):421-427

[93] Koh L, Ng D. Osteoporosis risk factor assessment and bone densitometry–current status and future trends. Annals of the Academy of Medicine, Singapore. 2002;**31**(1):37-42

[94] Chiappelli F. Osteoimmunopathology: Evidence-Based Perspectives from Molecular Biology to Systems Biology. Dordrecht: Springer; 2011

[95] NIH Consensus Development Panel on Osteoporosis Prevention, Diagnosis, and Therapy. Osteoporosis prevention, diagnosis, and therapy. JAMA. 2001;**285**(6):785-795

[96] Jarcho M. Calcium phosphate ceramics as hard tissue prosthetics. Clinical Orthopaedics and Related Research. 1981;**157**:259-278

[97] Hench LL et al. Bonding mechanisms at the interface of ceramic prosthetic materials. Journal of Biomedical Materials Research. 1971;**5**(6):117-141

[98] Jarcho M et al. Tissue, cellular and subcellular events at a bone-ceramic hydroxylapatite interface. Journal of Bioengineering. 1977;**1**(2):79-92

[99] Ducheyne P, De Groot K. In vivo surface activity of a hydroxyapatite alveolar bonesubstitute. Journal of Biomedical Materials Research. 1981;**15**(3):441-445

[100] LeGeros R et al. Biphasic calcium phosphate bioceramics: Preparation, properties and applications. Journal of Materials Science. Materials in Medicine. 2003;**14**(3):201-209

[101] Kitsugi T et al. Bone bonding behavior of three kinds of apatite containing glass ceramics. Journal of Biomedical Materials Research. 1986;**20**(9):1295-1307

[102] Kokubo T, Takadama H. How useful is SBF in predicting in vivo bone bioactivity? Biomaterials. 2006;**27**(15):2907-2915

[103] Kokubo T. Bioactive glass ceramics: Properties and applications. Biomaterials. 1991;**12**(2):155-163

[104] Bohner M, Lemaitre J. Can bioactivity be tested in vitro with SBF solution? Biomaterials. 2009;**30**(12):2175-2179

[105] Bertazzo S et al. Hydroxyapatite surface solubility and effect on cell adhesion. Colloids surf. B Biointerfaces. 2010;**78**(2):177-184

[106] Hench LL, Wilson J. Surface-active biomaterials. Science. 1984;**226**(4675):630-636

[107] Gross U, Schmitz HJ, Strunz V. Surface activities of bioactive glass, aluminum oxide, and titanium in a living environment. Annals of the New York Academy of Sciences. 1988;**523**(1):211-226

[108] Hoffman AS. Hydrogels for biomedical applications. Advanced Drug Delivery Reviews. 2002;**54**(1):3-12

[109] Driessens F et al. Effective formulations for the preparation of calcium phosphate bone cements. Journal of Materials Science. Materials in Medicine. 1994;**5**(3):164-170

[110] Gkioni K et al. Mineralization of hydrogels for bone regeneration. Tissue Engineering. Part B, Reviews. 2010;**16**(6):577-585

[111] Sherwood JK et al. A three-dimensional osteochondral composite scaffold for articular cartilage repair. Biomaterials. 2002;**23**(24):4739-4751

[112] Mikos AG et al. Preparation of poly(glycolic acid) bonded fiber structures for cell attachment and transplantation. Journal of Biomedical Materials Research. 1993;**27**(2):183-189

[113] Langer RS, Vacanti JP. Preparation of three-dimensional fibrous scaffold for attaching cells to produce vascularized tissue in vivo. Google Patents; 1998

[114] Ginty PJ et al. Mammalian cell survival and processing in supercritical CO_2. Proceedings of the National Academy of Sciences of the United States of America. 2006;**103**(19):7426-7431

[115] Stevens MM. Biomaterials for bone tissue engineering. Materials Today. 2008;**11**(5):18-25

[116] Tanahashi M, Matsuda T. Surface functional group dependence on apatite formation on self-assembled monolayers in a simulated body fluid. Journal of Biomedical Materials Research. 1997;**34**(3):305-315

[117] Stephansson SN, Byers BA, García AJ. Enhanced expression of the osteoblastic phenotype on substrates that modulate fibronectin conformation and integrin receptor binding. Biomaterials. 2002;**23**(12):2527-2534

[118] Thorwarth M et al. Bioactivation of an anorganic bone matrix by P-15 peptide for the promotion of early bone formation. Biomaterials. 2005;**26**(28):5648-5657

[119] Filmon R et al. Effects of negatively charged groups (carboxymethyl) on the calcification of poly(2-hydroxyethyl methacrylate). Biomaterials. 2002;**23**(14):3053-3059

[120] Song J, Malathong V, Bertozzi CR. Mineralization of synthetic polymer scaffolds: A bottom-up approach for the development of artificial bone. Journal of the American Chemical Society. 2005;**127**(10):3366-3372

Chapter 4

Proteins in Calcium Phosphates Biomineralization

Marta Kalka, Anna Zoglowek, Andrzej Ożyhar and Piotr Dobryczycki

Abstract

Biomineralization is a process of creating crystalline structures under biological control. The process takes place in hard tissues, such as bones, cartilages, and teeth. Biominerals are a combination of a crystal phase deposited onto an organic matrix. Inorganic components are mainly responsible for the biomineral hardness, while the organic matrix controls the shape, size, and polymorph of the crystals. Within the organic matrix, proteins exhibit a special biomineralization activity. Among them, one can distinguish insoluble collagen and soluble noncollagenous proteins. It is particularly noteworthy that noncollagenous proteins are intrinsically disordered proteins. High flexibility, acidic nature, and susceptibility to modifications make them especially adapted to the biomineralization control. This review paper is dedicated to the proteins which are involved in biomineralization of bones and teeth.

Keywords: biomineralization, bones, teeth, SIBLINGs, intrinsically disordered proteins, collagen

1. Introduction

Biomineralization is a process of formation of an inorganic solid within the biological system [1]. Biominerals are organic-inorganic composites, which fulfill various biological functions. In vertebrates, hard tissues provide body support, take part in tearing food, protect organs, and are reservoirs of calcium and phosphate. Understanding of molecular basis of biomineralization is essential to obtain new biomaterials.

Human hard tissues are formed of calcium orthophosphates [2]. Among them, most important are amorphous calcium phosphate, octacalcium phosphate, calcium hydrogenphosphate dihydrate, and calcium-deficient apatite and hydroxyapatite (HA) [3]. In organisms, calcium orthophosphates occur mainly in the form of poorly crystallized nonstoichiometric sodium-, magnesium-, and carbonate-containing HA—so-called biological apatite [2–5].

Biomineralization is a multistep process, which requires using the structures of extracellular matrix vesicles, numerous enzymes and glycoproteins. Due to strict control, biominerals differ from pure chemical minerals [6]. In contrast to geological minerals, biominerals are composite materials comprised of both inorganic and bioorganic components. The mineral constituent gives tissues hardness and resistance to mechanical damage. Stiffness of the tissue depends on the amount of inorganic components and organic phase [7]. About 70% of bone tissue is made up of mineral structure, while the rest is water and organic substances [1]. Moreover,

Protein name	Accession number	pI	Mass (kDa)	Overall percent disordered—PONDR
Osteopontin	AAA59974.1	4.4	~34	70%
Bone sialoprotein	AAA60549.1	4.1	~35	59%
Matrix extracellular phosphoglycoprotein	AAK70343.1	8.6	~58	64%
Dentin matrix protein-1	AAC51332.1	4.0	~56	90%
Dentin sialophosphoprotein (preproprotein)	NP_055023.2	3.6	~131	88%
Ameloblastin	AAG35772.1	4.9	~48	58%
Amelogenin	AAC21581.1	6.5	~22	65%

Bioinformatic predictions of a disordered structure in proteins were done with PONDR predictor (http://www.pondr.com/) [13].

Table 1.
Proteins involved in bone and teeth biomineralization.

having formed *in vivo* under well-controlled conditions, biomineral phases often have properties, such as shape, size, crystallinity, isotopic, and trace element compositions, quite unlike their inorganically formed counterparts [2].

The key factors, which determine the size, shape, internal structure and properties of biominerals, are proteins, which control the nucleation and growth of the crystals. Biomineralization involves protein-protein interactions and interactions between proteins and inorganic fraction. Among them, two major groups can be distinguished. Scaffold for a growing mineral phase provides insoluble collagen. In human bones, collagen makes up to 20–30% wt. Nucleation and crystal growth regulation are controlled by soluble, noncollagenous proteins. They are usually highly acidic [8, 9], undergo extensive posttranslational modifications [10], and frequently belong to the group of intrinsically disordered proteins (IDPs) [11]. IDPs are dynamic, flexible, heterogeneous populations of molecules without a well-defined folded structure. The highly charged character, along with the low content of hydrophobic amino acid residues, results in strong electrostatic repulsion and the lack of a well-packed hydrophobic core [12]. High content of IDPs' carboxyl and phosphate acidic groups involved in biomineralization results in high calcium binding capabilities. Examples of proteins engaged in bones and teeth biomineralization are shown in **Table 1**.

2. The basic division of proteins involved in the formation of calcium phosphate biominerals

Bone and teeth biomineralization takes place within the extracellular matrix (ECM), outside the cells. ECM is a highly organized and complex structure, unique to the specific organ type. It is an organic, noncellular fraction of mineralized tissues, composed of structural and functional proteins [14]. The content of the organic fraction varies depending on the type of tissue and constitutes 30% of bone, 20–25% of dentin, and only 0.5% of mature enamel. The basic division of proteins involved in the formation of calcium phosphate biominerals includes insoluble matrix like collagen and soluble noncollagenous proteins (NCPs).

2.1 Collagenous proteins of ECM

Among organic phase components, there are insoluble substances, which include structural macromolecules, creating a scaffold for the growing mineral.

They give shape to the resulting crystal and can act as initiators for the creation of crystallization roots. An example of such substance is collagen. Animal tissues primarily contain collagen type I or collagen type II [15]. Collagen is the major ECM molecule that self-assembles into cross-striated fibrils, provides support for cell growth, and is responsible for the mechanical resilience of connective tissues. The term "collagen" defines a whole family of glycoproteins. The most common motifs in the amino acid sequence of collagen are repeating sequence [Gly-X-Y]n, both with and without interruptions. The X and Y positions are occupied by proline and its hydroxylated form, hydroxyproline, respectively. The right-handed triple helix is formed from three left-handed polyproline α-chains of identical length, which gives collagen a unique quaternary structure [16]. Hydroxylation of the rest of the proline in collagen is necessary for the stabilization of the triple helix [17]. Collagen chains create interactions leading to the organization of chains in a four-stage array [18]. Collagen fibers are also cross-linked via lysyl residues [19]. Electron microscopy techniques allow to observe successive regions of overlap and gaps, resulting from the mutual arrangement of collagen fibers [20].

2.2 Noncollagenous proteins (NCPs) of extracellular matrix (ECM)

While structural collagen is the main protein of ECM, NCPs are important for the regulation of biomineralization. NCPs are usually acidic, capable of binding large amounts of calcium ions. Proteins, which directly influence biomineralization, act as nucleators and regulators of crystal growth and orientation [21]. NCPs are frequently posttranslationally modified, e.g., phosphorylated, glycosylated, or proteolytically processed. A large number of noncollagenous proteins have disordered secondary and tertiary structures and belong to a group of intrinsically disordered proteins [11].

3. Intrinsically disordered proteins in calcium phosphate biomineralization

IDPs are a group of proteins that have gained a special interest of researchers for over 20 years [12, 22, 23]. Their discovery was especially challenging for traditional protein structure paradigm, stating that protein function depends on a fixed tertiary structure. The IDP group includes proteins with altered secondary and tertiary structures from total random coil structure to some intrinsically disordered regions. They are characterized by considerable plasticity, flexibility, and high conformational dynamics. IDP features are high net charge and low hydrophobicity, which is not conducive to the formation of the hydrophobic core. They often bind low molecular ligands and macromolecules such as ions and proteins. By interacting with ligands, IDPs can undergo local or global structuring. They can also fulfill their functions while remaining completely disordered. Other characteristics of this group are multiple amino acid repeats and susceptibility to posttranslational modifications. An open structure and lack of a packed core increase the availability of potential phosphorylation, glycosylation, and proteolytic cleavage sites [24].

Extended, flexible structure, highly acidic nature, susceptibility to modifications and especially the ability to interact with many different partners make these proteins particularly adapted to the biomineralization control. Many NCPs belong to IDP group. Family of small integrin-binding ligand N-linked glycoprotein (SIBLING) is the example of IDPs engaged in HA formation and is presented below. In addition to SIBLINGs, enamel matrix proteins such as ameloblastin, amelotin, and enamelin also have disordered structures [25]. More information about the proteins involved in HA biomineralization belonging to IDPs can be found in reviews [11, 25, 26].

3.1 SIBLINGs

An example of NCPs is the family of small integrin-binding ligand N-linked glycoprotein (SIBLING). It is a group of proteins identified in bone, dentine, and cementum, which includes osteopontin (OPN), bone sialoprotein (BSP), dentin matrix protein-1 (DMP1), dentin sialophosphoprotein (DSPP), and matrix extracellular phosphoglycoprotein (MEPE) [27]. It is believed that they all evolved as a result of gene duplication from extracellular calcium-binding phosphoprotein family. Although proteins differ in composition of amino-acid sequence, SIBLINGs share common features [28]. Their genes are located in a 375 kb region on the 4q21 human and 5q mouse chromosomes. They display a similar exon structure. Their sequence contains an Arg-Gly-Asp (RGD) motif, which mediates cell attachment/signaling by binding to cell-surface integrins [28]. SIBLINGs belong to IDP group and are highly acidic. They share numerous sequence repeats, which are often observed in the case of IDPs [26]. Particularly important for their biomineralization activity may be acidic serine- and aspartate-rich motifs (ASARM), which are involved in the phosphate administration [29, 30]. SILBINGs are frequently posttranslationally medicated. Some of their biological functions depend on phosphorylation, glycosylation, proteolytic processing, sulfonation, or transglutaminase cross linking [31].

3.1.1 Osteopontin

Osteopontin (OPN), also known as secreted phosphoprotein 1 (SSP1), is a highly phosphorylated and glycosylated sialoprotein. It is a multifunctional protein expressed by several cell types including osteoblasts, osteocytes, as well as osteoclasts or odontoblasts [32]. OPN undergoes many posttranslational modifications such as phosphorylation, glycosylation, sulfonation, or proteolytic processing, and modifications vary depending on a protein role and localization [31]. In bones and dentin, osteopontin is located at the site of biomineral formation. Synthesis of OPN is stimulated by calcitriol, physiologically active form of vitamin D, which is known as a trigger for bone destruction and remodeling [33, 34].

Osteopontin was originally identified as a bridge between the cell and HA in ECM of bone [35]. It is known that OPN increases the adhesion of bone cells by concentrating in mineralized collagen matrix during bone formation and remodeling [36, 37]. The protein is highly produced by developing osteoblasts and osteoclasts, and it has been shown to regulate both cell type adhesions.

Another role of ONP during biomineralization is modulation of osteoclastic function. OPN binding to integrin $\alpha v\beta 3$ is essential for regulation of osteoclastic activity and is necessary in the formation of a sealing zone [34]. Through CD44-associated cell signaling, OPN stimulates osteoclast migration [38].

As a highly acidic phosphoprotein, OPN binds to the surface of hydroxyapatite crystals through the electrostatic interactions between crystals and carboxyl and phosphate groups. Furthermore, disordered structure of OPN can promote the binding of calcium ions. The results of *in vitro* and *in vivo* studies suggest that osteopontin has an inhibitory effect on HA formation and growth [39, 40]. *OPN* knockout mice display increased mineral content and size [41]. Therefore, it is suggested that OPN is one of the proteins whose role is to prevent crystal formation in soft tissues. Peptides of OPN obtained after proteolytic processing differ in a biomineralization effect. ASARM peptide of OPN binds to HA crystals and consequently inhibits ECM matrix mineralization [42]. Its inhibition activity is dependent on serine residues' phosphorylation. N- and C-terminal fragments of OPN from milk promote *de novo* HA formation, but at the same time, the central fragment showed an inhibitory effect [43].

3.1.2 Bone sialoprotein

Bone sialoprotein (BSP) is one of the most abundant NCPs of bone. In contrast to OPN, BSP is only localized in tissues that undergo mineralization: bone, dentin and mineralizing cartilage, and cementum [44]. The protein is produced by osteoblasts, osteocytes, osteoclasts, as well as by chondrocytes. Disordered structure of BSP was shown by NMR, CD, and SAXS studies [28, 45–47]. BSP is highly glycosylated especially at C-terminal fragment, and to a lesser extent also phosphorylated and sulfonated [31]. N-terminal region of the protein contains collagen-binding motif. *In vivo* and *in vitro* studies indicated that BSP can be involved in initial stages of hydroxyapatite biomineralization [48]. BSP is intensely expressed by osteoblastic cells in sites of primary bone formation. The protein is present at mineralizing boundaries of bone, dentin, and calcifying cartilage tissues. On the other hand, BSP increases osteoclastogenesis, and in that way, it can initiate bone remodeling. *BSP* knockout mice are characterized by short, hypomineralized bones with high trabecular bone mass [49, 50]. Overexpression of BSP leads to dwarfism, decreased bone mineral density, and decreased trabecular bone volume [51].

It was shown that BSP has high affinity to hydroxyapatite and acts as a *de novo* nucleator of HA crystals *in vitro* [52]. The protein also binds to type I collagen, and by interaction with collagen, BSP could regulate HA nucleation [53]. Moreover, control of mineral formation is highly associated with the state of phosphorylation, sulfonation, and glycosylation of BSP [27, 54, 55].

When creating skull bones, including the mandible, alveoli, and skull vault, the osteoprogenitor cells secrete a collagen extracellular matrix—osteoid. Skull bones are formed by intracerebral ossification. BSP affects biomineralization of the matrix in the ossification process, hence, the formation, shaping, and growth of hydroxyapatite crystals. In the absence of BSP, bone formation is delayed and osteoblast activity is impaired [56].

3.1.3 Matrix extracellular phosphoglycoprotein

Matrix extracellular phosphoglycoprotein (MEPE) is located in mineralized tissues as bone and dentin, but was found also in nonmineralized organs. The protein is primarily expressed by osteoblasts and osteocytes that are embedded within the mineralized matrix in bone and by odontoblasts during odontogenesis [57]. Similar to osteopontin, MEPE seems to be a multifunctional ECM protein. Results of *in vivo* and *in vitro* studies suggest that MEPE is important for bone mineralization, Pi homeostasis and cell attachment [58–60]. The protein was originally identified as interacting with PHEX. PHEX by binding to MEPE protects the protein from proteolytic cleavage by cathepsin B. Cleavage by cathepsin B leads to the release of the ASRAM peptide. The ASRAM motif is a functional domain of MEPE responsible for its inhibitory activity [61]. The peptide may be responsible for phosphate and calcium regulation during the mineralization process. ASARM peptide inhibits hydroxyapatite mineralization by binding free calcium with high avidity, and their inhibitory effect is highly dependent on serine residues phosphorylation. AC100, another MEPE fragment, containing the integrin binding (RGD) and glycosaminoglycan-attachment (SGDG) motifs stimulates new bone formation *in vitro* and *in vivo* [62].

3.1.4 Dentin matrix protein-1

Dentin matrix protein-1 (DMP1) was the first molecule identified by cloning from the dentin matrix. Besides dentin, the protein is located also in bone and cementum as well as in other nonmineralized tissues [63–65]. It is remarkably acidic and is a hydrophilic protein with many serine, aspartate, and glutamate residues. The protein is characterized by disordered structure that can aggregate in the presence of calcium ions [66].

It was shown that DMP1 is essential for both odontoblasts and osteoblasts maturation. The protein controls odontogenesis, osteogenesis, and Pi homeostasis. The effect of DMP1 on HA biomineralization depends on posttranslational modifications. It was shown that phosphorylated DMP1 inhibits HA formation and growth, while the dephosphorylated form acts as a HA nucleator [67, 68].

It was demonstrated that DMP1 is proteolytically processed into N-terminal 37-K fragment and 57-K fragment from the C-terminal region, and it is likely that full-length DMP1 is a precursor to these functional fragments [69]. It seems that the 57-K fragment plays a key role in the biomineralization process. The 57-K fragment contains 41 phosphate groups, while the 37-K fragment has only 12 phosphate groups. In addition, the 57-K fragment contains many functional sequences and domains such as the RGD motif [70], the ASARM peptide [71], and the peptide functioning as nucleator [72, 73]. In contrast to the full-length DMP1, the 57-K fragment promoted HA nucleation and growth [66]. Furthermore, it was shown that DMP1 N-terminal fragment is able to stabilize the metastable amorphous calcium phosphate. Due to high aspartic acid residual content, the 37-K fragment can bind calcium ions very strongly favoring formation and stabilization of the amorphous nuclei [72]. It was shown that the DMP1 57-K fragment also controls calcium carbonate mineralization *in vitro* [74].

3.1.5 Dentin sialophosphoprotein

Dentin sialophosphoprotein (DSPP) was discovered by cDNA cloning using a mouse odontoblast cDNA library and was the first believed to be connected only with dentin. Subsequent studies have revealed that the protein is expressed also in bone, cementum, and some nonmineralized tissues [75, 76]. Analysis of *DSPP* knockout mice and mutation in the *DSPP* gene indicated a special role of protein in dentin mineralization. In humans, a mutation in the *DSPP* gene causes dentin hypomineralization and significant tooth decay, named dentinogenesis imperfecta [77]. Additionally, studies of *DSPP*-null mice suggest that DSPP is crucial in the initial mineralization of bone as well as in the remodeling of the skeleton and therefore on bone turnover [78].

It is suggested that DSPP is a precursor protein activated after proteolytic processing. Cleavage of DSPP by zinc metalloproteinase bone morphogenetic protein-1 (BPM1) results in two protein fragments: dentin phosphoprotein (phosphophoryn) (DPP) and dentin sialoprotein (DSP) [79]. Subsequently, cleavage of C-terminal of DSP by matrix metalloproteinase 2 (MMP2) and MMP20 leads to the release of the third fragment called dentin glycoprotein (DGP), which has a strong affinity to hydroxyapatite [80].

Highly acidic DPP is the most abundant NCP of ECM in dentin. Amino acid sequence of DPP is composed mainly of aspartic acid and serine residues, whereas ~80% of serine residues are phosphorylated and phosphorylation is crucial to its function. The protein is also glycosylated [81] and contains ASARM peptide. Due to its amino acid composition and phosphorylation, DPP binds large amounts of calcium ions with high affinity. The protein is involved in nucleation and control of the formation and growth of HA crystals during dentin mineralization [82, 83]. DPP binds to collagen fibrils and is present in front of dentin mineralization [84, 85]. There is a hypothesis that in front of mineralization, DPP is promoting the formation of initial apatite crystals [86]. When predentin is converted to dentin, DPP with other proteins binds to the growing HA faces and inhibits or slows down crystal growth [86–88].

4. Other proteins specific to teeth or bones

4.1 Tooth

Dentin, enamel, and cementum are calcified tissues and are major components of teeth. Dentin is usually covered by enamel on the crown and cementum on the root and surrounds the entire pulp. About 70% (wt) of dentin consists of the mineral phase, 20% (wt) is an organic material, and 10% (wt) is water. Enamel is the hardest substance in the human body and contains 96% (wt) of mineral phase. Enamel and dentin are created by two layers, namely odontoblasts and ameloblasts. Dentin is characterized by a high content of collagen I comprising ~90% of the dentin organic matrix [89].

4.1.1 Enamel

The organic part of enamel consists mainly of hydrophobic proteins, known as amelogenins and anionic proteins, which include ameloblastin, enamelin, and tuftelin [90]. Enamel also contains sialophosphoproteins and enamel proteases such as matrix-20 metalloproteinase (MMP-20, matrix metalloproteinase-20, also known as enamelysin) and kallikrein 4 (KLK4). At the end of the crystallization, enamel loses almost entire organic matrix, which is degraded and replaced by hydroxyapatite crystals [90]. Residues, small peptides, and amino acids account for only 1% of enamel and do not resemble the original matrix [90].

Amelogenins are a group of heterogeneous proteins present in odontoblasts and identified in enamel of maturing teeth. They are the main proteins of the organic enamel matrix, constituting over 90% of the pool of proteins secreted by ameloblasts into the intercellular matrix during the formation of enamel. These are hydrophobic proteins rich in proline, glutamine, leucine, and histidine. Amelogenins isolated from various organisms are characterized by high homology [91]. Amelogenins belong to IDP groups; conformational analysis by CD and NMR spectroscopy showed that recombinant porcine amelogenin rP172 exists in an extended, unfolded state in the monomeric form [92]. The extended, labile conformation of rP172 amelogenin is compatible with known functions of amelogenin in enamel biomineralization, i.e., self-assembly, associations with other enamel matrix proteins and with calcium phosphate biominerals, and interaction with cell receptors [92].

Amelogenin controls the organization and growth of enamel crystals, and its presence is critical for normal enamel formation. Defects in amelogenin sequence lead to defective enamel crystal formation and organization [93].

Ameloblastin, also known as amelin, is a tooth-specific glycoprotein. Unlike amelogenin, ameloblastin is located close to the cell surface [94]. Ameloblastin accounts for about 5–10% of all proteins present in the enamel, and it is the second most abundant protein among the intercellular matrix of the enamel [95]. Ameloblastin is also synthesized in dentin and cement, but its role in these tissues is not determined. Ameloblastin is a molecule of cell adhesion required to maintain a diversified state of ameloblasts [96]. Bioinformatic analyzes and molecular modeling of the protein structure suggest that it may belong to IDPs [97], which was confirmed by the CD spectra of recombinant mouse protein [98].

The N-terminal fragment of ameloblastin comprising the self-assembly motif was shown to colocalize with amelogenin across the entire growing enamel, indicating the role of the two proteins in the organization of the linear growth of HA crystallites [99–101]. Both proteins self-assemble into higher order structures from monomeric subunits, similar to type I collagen, the predominant matrix protein of bones, and dentin [102, 103]. Full-length amelogenin undergoes hierarchical stepwise assembly, first forming oligomers, which in turn assemble into higher order structures and stabilize mineral prenucleation clusters, and organize them into parallel arrays of linear chains, yielding the formation of crystallite bundles [102]. There is a hypothesis that the higher order structures of the self-assembled ameloblastin (or most likely its N-terminal moiety) contribute to the oriented growth of the linear chains of amelogenin in the 3D space [104].

4.1.2 Dentin

Dentin forms a support for hard tissues of a tooth. It covers both crown and root structures and is responsible for its shape. The formation of dentin is closely related to bone. Both bone and dentin are composed of a collagenous matrix and a mineral phase with hydroxyapatite plate-like crystals. Mineralization of dentin and bone extracellular matrix is initiated with the aid of matrix vesicles and later involves secretion of families of a specialized matrix protein [105]. In contrast to bones, dentin is not remodeled and does not participate in the regulation of calcium and phosphate metabolism. The tooth formation takes place throughout the lifetime of a tooth. Unlike enamel, dentin mineral content increases with age due to continuous mineral deposition either as physiological secondary dentin or as tertiary dentin following injury [105].

Dentin is composed of a collagen I, which accounts for about 92% of organic components. In addition to collagen, there is also a group of noncollagen proteins, which include dentin-specific phosphoryls (DPPs) and dentin sialoproteins (DSP), sialic acid, osteopontin, dentin matrix proteins (DMP1, DMP2, DMP3), BSP, acidic bone-75 glycoprotein, osteonectin, and proteins rich in γ-carboxyglutamic acid. Among organic components should also be mentioned proteoglycans, growth factors, phospholipids, and enzymes [106, 107]. Inorganic dentin components make up to about 70% of the tissue mass, making the dentin harder than the bone. The mass of a single hydroxyapatite crystal is about 10 times greater than in bone, but many times less than in enamel, and its size is about $35 \times 10 \times 100$ nm [108].

4.1.3 Cementum

Cementum protects the dental root dentin with a very thin layer. In many respects, it is very unique: it is not vascularized and is not innervated, it does not undergo constant remodeling like a bone, but it grows all the time. Cellular and acellular cement are distinguished based on the presence of cementocytes in its structure. The structure and composition of cement resembles bone tissue. Inorganic constituents make up to about 65% of tissue mass and consist mainly of hydroxyapatite crystals [109, 110].

Cementum provides contact between roots of the teeth and periodontium ligaments [111]. hrCEMP1 (recombinant human cementum protein 1) is an isolated form of cementum human β-sheet protein. It has been showed that hrCEMP1 forms clusters of 6.5 nm diameter [111]. hrCEMP1 is an inducer of specific biomineralization proteins. It stimulates the differentiation of osteoblasts and cementoblasts. It is suggested that hrCEMP1 plays a significant role in octacalcium phosphate biomineralization and helps its binding to hydroxyapatite even without posttranslational modifications. X-ray diffraction measurements showed that hrCEMP1 is an inducer of polymorphic crystals [111].

4.2 Bone

The matrix of bone consists of type I collagen molecule self-assembled into a triple helix consisting of two α1 and one α2 chains (ca. 300 nm in length and 1.5 nm in diameter) and hydroxyapatite ($Ca_{10}(PO_4)_6(OH)_2$) nanocrystals (plate-shaped, 1.5–4 nm in thickness) deposited along collagen fibrils [112, 113]. The collagen fiber does not undergo spontaneous mineralization in the presence of phosphate and calcium ions [20], and the involvement of noncollagen proteins is necessary. The majority of these proteins belong to the SIBLING family described above [27].

Bones and dentin are characterized by a similar composition and the mechanism of their formation [112]. Their organic matrix consists of type I collagen and the mineral matrix built of hydroxyapatite. Osteoblasts and odontoblasts, cells involved in osteogenesis and dentinogenesis, first secrete a nonmineralized matrix—a bone osteoid, and in case of dentin—predentin and then with fibers of collagen I form matrix for biomineralization [112]. Calcium and phosphate ions are dislocated from the vascular network into the mineralization matrix. Osteoid and predentin contain numerous noncollagenous proteins called NCPs. On the basis of mutation experiments and suppression of NCP genes, it is suggested that they stimulate the nucleation and growth of hydroxyapatite crystals [112].

Bones and dentin differ significantly. Bone is a dynamic tissue because it is constantly remodeled, while dentin is a rather static tissue [112]. Osteoblasts produce matrix components as a result of controlling growth factors to form a bone. Hydroxyapatite crystals grow above vesicles, thus creating mineralization centers. Bones have a lot of vessels, and they store calcium ions. The reason for bone resorption is the action of lysosomal enzymes produced by osteoclasts and lowering the pH in extracellular bone matrix.

Osteoclasts are responsible for the attachment, bone degradation, and subsequent tissue resorption. It was found that this is possible due to the strong polarization of osteoclasts. They have both frontal and posterior abdominal polarization. On the underside of the osteoclasts, there are structures responsible for the degradation of the mineral surface. The adhesion zone is located in the sealing zone [114]—a ring made of actin and podosomes in which extracellular resorption pit is segregated and is used to retain heavy substances (minerals) formed during bone resorption near the bone surface [43, 114]. Podosomes are dynamic structures built of actin. They perform important functions, including adhesion, destruction of the bone and its matrix, and recognition of the appropriate medium [115]. Podosomes contain a large number of V-ATPase molecules that pump protons and secrete proteases, which promote bone degradation [114].

Acknowledgements

This work was supported by the National Science Center (Poland) [UMO-2015/19/B/ST10/02148] and in a part by statutory activity subsidy from the Polish Ministry of Science and High Education for the Faculty of Chemistry of Wroclaw University of Science and Technology.

Conflict of interest

Authors declare no conflict of interest.

Author details

Marta Kalka, Anna Zoglowek, Andrzej Ożyhar and Piotr Dobryczycki*
Department of Biochemistry, Faculty of Chemistry, Wrocław University of Science and Technology, Wrocław, Poland

*Address all correspondence to: piotr.dobryszycki@pwr.edu.pl

References

[1] Lowenstam HA, Weiner S. On Biomineralization. New York: Oxford University Press; 1989. 324 p

[2] Dorozhkin SV. Calcium orthophosphates occurrence, properties, biomineralization, pathological calcification and biomimetic applications. Biomatter. 2011;**1**:121-164

[3] Dorozhkin SV. Calcium orthophosphates ($CaPO_4$): Occurrence and properties. Progress in Biomaterials. 2016;**5**:9-70

[4] Grynpas MD, Omelon S. Transient precursor strategy or very small biological apatite crystals? Bone. 2007;**4**:162-164

[5] Young RA. Biological apatite vs hydroxyapatite at the atomic level. Clinical Orthopaedics and Related Research. 1975;**113**:249-262

[6] Wopenka B, Pasteris JD. A mineralogical perspective on the apatite in bone. Materials Science and Engineering. 2005;**C 25**:131-143

[7] Hołubowicz R, Porębska A, Poznar M, Różycka M, Dobryszycki P. Biomineralization-precision of shape, structure and properties controlled by proteins. Postepy Biochemii. 2015;**61**:364-380

[8] Marin F, Luquet G. Unusually acidic proteins in biomineralization. In: Handbook of Biomineralization: Biological Aspects and Structure Formation. USA: Wiley; 2008. DOI: 10.1002/9783527619443.ch16

[9] Gorski JP. Acidic phosphoproteins from bone matrix: A structural rationalization of their role in biomineralization. Calcified Tissue International. 1992;**50**:391-396

[10] Jain A, Karadag A, Fohr B, Fisher LW, Fedarko NS. Three SIBLINGs (small integrin-binding ligand, N-linked glycoproteins) enhance factor H's cofactor activity enabling MCP-like cellular evasion of complement-mediated attack. The Journal of Biological Chemistry. 2002;**277**:13700-13708

[11] Wojtas M, Dobryszycki P, Ożyhar A. Intrinsically disordered proteins in biomineralization. In: Advanced Topics in Biomineralization. UK: IntechOpen; 2012. pp. 3-32. DOI: 10.5772/31121

[12] Uversky VN, Gillespie JR, Fink AL. Why are "natively unfolded" proteins unstructured under physiologic conditions? Proteins. 2000;**41**:415-427

[13] Romero P, Obradovic Z, Li X, Garner EC, Brown CJ, Dunker AK. Sequence complexity of disordered protein. Proteins. 2001;**42**:38-48

[14] Ravindran S, George A. Multifunctional ECM proteins in bone and teeth. Experimental Cell Research. 2014;**325**(2):148-154

[15] Alberts B, Johnson A, Lewis J, et al. Molecular biology of the cell. In: The Extracellular Matrix of Animals. Available from: https://www.ncbi.nlm.nih.gov/books/NBK26810/. 4th ed. New York: Garland Science; 2002

[16] Brodsky B, Persikov AV. Molecular structure of the collagen triple helix. Advances in Protein Chemistry. 2005;**70**:301-339

[17] Van der Rest M, Garrone R. Collagen family of proteins. The FASEB Journal. 1991;**13**:2814-2823

[18] Olszta MJ, Cheng X, Jee SS, Kumar R, Kim Y, Kaufman MJ, et al. Bone structure and formation, a new perspective. Materials Science and Engineering R. 2007;**58**:77-116

[19] Knott L, Bailey AJ. Collagen cross-links in mineralizing tissues, a review of their chemistry, function, and clinical relevance. Bone. 1998;**22**:181-187

[20] Nudelman F, Lausch AJ, Sommerdijk NA, Sone ED. In vitro models of collagen biomineralization. Journal of Structural Biology. 2013;**183**:258-269

[21] Boskey AL. Biomineralization: An overview. Connective Tissue Research. 2003;**44**:5-9

[22] Wright PE, Dyson HJ. Intrinsically unstructured proteins, re-assessing the protein structure-function paradigm. Journal of Molecular Biology. 1999;**293**:321-331

[23] Tompa P. Intrinsically unstructured proteins. Trends in Biochemical Sciences. 2002;**27**:527-533

[24] Iakoucheva LM, Radivojac P, Brown CJ, O'Connor TR, Sikes JG, Obradovic Z, et al. The importance of intrinsic disorder for protein phosphorylation. Nucleic Acids Research. 2004;**32**:1037-1049

[25] Kalmar L, Homola D, Varga D, Tompa P. Structural disorder in proteins brings order to crystal growth in biomineralization. Bone. 2012;**51**:528-534

[26] Boskey AL, Villarreal-Ramirez E. Intrinsically disordered proteins and biomineralization. Matrix Biology. 2016;**52-54**:43-59

[27] Staines KA, MacRae VE, Farquharson C. The importance of the SIBLING family of proteins on skeletal mineralisation and bone remodelling. Journal of Endocrinology. 2012;**214**:241-255

[28] Fisher LW, Torchia DA, Fohr B, Young MF, Fedarko NS. Flexible structures of SIBLING proteins, bone sialoprotein, and osteopontin. Biochemical and Biophysical Research Communications. 2001;**280**:460-465

[29] Rowe PS, Kumagai Y, Gutierrez G, Garrett IR, Blacher R, Rosen D, et al. MEPE has the properties of an osteoblastic phosphatonin and minhibin. Bone. 2004;**34**:303-319

[30] David V, Martin AC, Hedge AM, Drezner MK, Rowe PS. ASARM peptides: PHEX-dependent, independent regulation of serum phosphate. American Journal of Physiology. Renal Physiology. 2010;**300**:F783-F791

[31] Qin C, Baba O, Butler WT. Post-translational modification of SIBLING proteins and their roles in osteogenesis and dentinogenesis. Critical Reviews in Oral Biology and Medicine. 2004;**15**:126-136

[32] Icer MA, Gezmen-Karadag M. The multiple functions and mechanisms of osteopontin. Clinical Biochemistry. 2018;**59**:17-24

[33] Noda M, Vogel RL, Craig AM, Prahl J, DeLuca HF, Denhardt DT. Identification of a DNA sequence responsible for binding of the 1,25-dihydroxyvitamin D3 receptor and 1,25-dihydroxyvitamin D3 enhancement of mouse secreted phosphoprotein 1 (SPP-1 or osteopontin) gene expression. Proceedings of the National Academy of Sciences of the United States of America. 1990;**87**:9995-9999

[34] Reinholt FP, Hultenby K, Oldberg A, Heinegård D. Osteopontin—A possible anchor of osteoclasts to bone. Proceedings of the National Academy of Sciences. 1990;**87**:4473-4475

[35] Sodek J, Ganss B, McKee MD. Osteopontin. Critical Reviews in Oral Biology and Medicine. 2000;**11**:279-303

[36] Giachelli CM, Steitz S. Osteopontin: A versatile regulator of inflammation and biomineralization. Matrix Biology. 2000;**19**:615-622

[37] McKee M, Nanci A. Osteopontin at mineralized tissue interfaces in bone, teeth, and osseointegrated implants: Ultrastructural distribution and implications for mineralized tissue formation, turnover, and repair. Microscopy Research and Technique. 1996;**33**:141-164

[38] Chellaiah M, Hruska K. The integrin αVβ3 and CD44 regulate the actions of osteopontin on osteoclast motility. Calcified Tissue International. 2002;**72**:97-205

[39] Boskey AL, Maresca M, Ullrich W, Doty SB, Butler WT, Prince CW. Osteopontin-hydroxyapatite interactions in vitro: Inhibition of hydroxyapatite formation and growth in a gelatin-gel. Bone and Mineral. 1993;**22**:147-159

[40] Hunter GK. Role of osteopontin in modulation of hydroxyapatite formation. Calcified Tissue International. 2013;**93**:348-354

[41] Boskey AL, Spevak L, Paschalis E, Doty SB, McKee MD. Osteopontin deficiency increases mineral content and mineral crystallinity in mouse bone. Calcified Tissue International. 2002;**71**:145-154

[42] Addison WN, Masica DL, Gray JJ, McKee MD. Phosphorylation-dependent inhibition of mineralization by osteopontin ASARM peptides is regulated by PHEX cleavage. Journal of Bone and Mineral Research. 2010;**25**:695-705

[43] Boskey AL, Christensen B, Taleb H, Sorensen ES. Post-translational modification of osteopontin: Effects on in vitro hydroxyapatite formation and growth. Biochemical and Biophysical Research Communications. 2012;**419**:333-338

[44] Bianco P, Fisher LW, Young MF, Termine JD, Robey PG. Expression of bone sialoprotein (BSP) in developing human tissues. Calcified Tissue International. 1991;**49**:421-426

[45] Tye CE, Rattray KR, Warner KJ, Gordon JA, Sodek J, Hunter GK, et al. Delineation of the hydroxyapatite-nucleating domains of bone sialoprotein. The Journal of Biological Chemistry. 2003;**278**:7949-7955

[46] Tye CE, Hunter GK, Goldberg HA. Identification of the type I collagen-binding domain of bone sialoprotein and characterization of the mechanism of interaction. The Journal of Biological Chemistry. 2005;**280**:13487-13492

[47] Wuttke M, Muller S, Nitsche DP, Paulsson M, Hanisch FG, Maurer P. Structural characterization of human recombinant and bone-derived bone sialoprotein. Functional implications for cell attachment and hydroxyapatite binding. The Journal of Biological Chemistry. 2001;**276**:36839-36848

[48] Ganss B, Kim RH, Sodek J. Bone sialoprotein. Critical Reviews in Oral Biology and Medicine. 1999;**10**:79-98

[49] Malaval L, Wade-Gueye NM, Boudiffa M, Fei J, Zirngibl R, et al. Bonesialoprotein plays a functional role in bone formation and osteoclastogenesis. The Journal of Experimental Medicine. 2008;**205**:1145-1153

[50] Bouleftour W, Boudiffa M, Wade-Gueye NM, Bouët G, Cardelli M, Laroche N, et al. Skeletal development of mice lacking bone sialoprotein (BSP)—Impairment of long bone growth and progressive establishment of high trabecular bone mass. PLoS One. 2014;**9**:e95144

[51] Valverde P, Zhang J, Fix A, Zhu J, Ma W, Tu Q, et al. Overexpression of bone sialoprotein leads to an uncoupling of bone formation and bone resorption in mice. Journal of Bone and Mineral Research. 2008;**23**:1775-1788

[52] Hunter GK, Goldberg HA. Nucleation of hydroxyapatite by bone sialoprotein. Proceedings of the National Academy of Sciences. 1993;**90**:8562-8565

[53] Baht GS, Hunter GK, Goldberg HA. Bone sialoprotein-collagen interactionpromotes hydroxyapatite nucleation. Matrix Biology. 2008;**27**:600-608

[54] George A, Veis A. Phosphorylated proteins and control over apatite nucleation, crystal growth, and inhibition. Chemical Reviews. 2008;**108**:4670-4693

[55] Baht GS, O'Young J, Borovina A, Chen H, Tye CE, Karttunen M, et al. Phosphorylation of Ser136 is critical for potent bone sialoprotein-mediated nucleation of hydroxyapatite crystals. The Biochemical Journal. 2010;**428**:385-395

[56] Foster BL, Ao M, Willoughby C, Soenjaya Y, Holm E, Lukashova L, et al. Mineralization defects in cementum and craniofacial bone from loss of bone sialoprotein. Bone. 2015;**78**:150-164

[57] Nampei A, Hashimoto J, Hayashida K, Tsuboi H, Shi K, Tsuji I, et al. Matrix extracellular phosphoglycoprotein (MEPE) is highly expressed in osteocytes in human bone. Journal of Bone and Mineral Metabolism. 2004;**22**:176-184

[58] Gowen LC, Petersen DN, Mansolf AL, Qi H, Stock JL, et al. Targeted disruption of the osteoblast/osteocyte factor 45 gene (OF45) results in increased bone formation and bone mass. Journal of Biological Chemistry. 2003;**278**:1998-2007

[59] Quarles LD. FGF23, PHEX, and MEPE regulation of phosphate homeostasis and skeletal mineralization. American Journal of Physiology. Endocrinology and Metabolism. 2003;**285**:1-9

[60] Dobbie H, Unwin RJ, Faria NJ, Shirley DG. Matrix extracellular phosphoglycoprotein causes phosphaturia in rats by inhibiting tubular phosphate reabsorption. Nephrology, Dialysis, Transplantation. 2008;**23**:730-733

[61] Addison WN, Nakano Y, Loisel T, Crine P, McKee MD. MEPE-ASARM peptides control extracellular matrix mineralization by binding to hydroxyapatite: An inhibition regulated by PHEX cleavage of ASARM. Journal of Bone and Mineral Research. 2008;**23**:1638-1649

[62] Hayashibara T, Hiraga T, Yi B, Nomizu M, Kumagai Y, Nishimura R, et al. A synthetic peptide fragment of human MEPE stimulates new bone formation in vitro and in vivo. Journal of Bone and Mineral Research. 2004;**19**:455-462

[63] George A, Sabsay B, Simonian PA, Veis A. Characterization of a novel dentin matrix acidic phosphoprotein, implications for induction of biomineralization. Journal of Biological Chemistry. 1993;**268**:12624-12630

[64] MacDougall M, Gu TT, Luan X, Simmons D, Chen J. Identification of a novel isoform of mouse dentin matrix protein 1: Spatial expression in mineralized tissues. Journal of Bone and Mineral Research. 1998;**13**:422-431

[65] Sun Y, Chen L, Ma S, Zhou J, Zhang H, Feng JQ, et al. Roles of DMP1 processing in osteogenesis, dentinogenesis and chondrogenesis. Cells, Tissues, Organs. 2011;**194**:199-204

[66] Gericke A, Qin C, Sun Y, Redfern R, Redfern D, Fujimoto Y, et al. Different

forms of DMP1 play distinct roles in mineralization. Journal of Dental Research. 2010;**89**:355-359

[67] He G, Dahl T, Veis A, George A. Dentin matrix protein 1 initiates hydroxyapatite formation in vitro. Connective Tissue Research. 2003;**44**:240-245

[68] Tartaix PH, Doulaverakis M, George A, Fisher LW, Butler WT, Qin C, et al. In vitro effects of dentin matrix protein-1 on hydroxyapatite formation provide insights into in vivo functions. Journal of Biological Chemistry. 2004;**279**:18115-18120

[69] Qin C, Brunn JC, Cook RG, Orkiszewski RS, Malone JP, Veis A, et al. Evidence for the proteolytic processing of dentin matrix protein 1. Identification and characterization of processed fragments and cleavage sites. Journal of Biological Chemistry. 2003;**278**:34700-34708

[70] Kulkarni GV, Chen B, Malone JP, Sampath Narayanan A, George A. Promotion of selective cell attachment by the RGD sequence in dentine matrix protein 1. Archives of Oral Biology. 2000;**45**:475-484

[71] Rowe PS. The wrickkened pathways of FGF23, MEPE and PHEX. Critical Reviews in Oral Biology and Medicine. 2004;**15**:264-281

[72] Gajjeraman S, Narayanan K, Hao J, Qin C, George A. Matrix macromolecules in hard tissues control the nucleation and hierarchical assembly of hydroxyapatite. The Journal of Biological Chemistry. 2007;**282**:1193-1204

[73] Lu Y, Qin C, Xie Y, Bonewald LF, Feng JQ. Studies of the DMP1 57-kDa functional domain both in vivo and in vitro. Cells, Tissues, Organs. 2008;**189**:175-185

[74] Porębska A, Ożyhar A, Dobryszycki P. Intrinsically disordered recombinant 57K fragment of human DMP1 influences the in vitro crystallization of $CaCO_3$. Protein Science. 2015;**24**(Suppl 1):137-138. DOI: 10.1002/pro.2823

[75] Qin C, Brunn JC, Cadena E, Ridall A, Tsujigiwa H, Nagatsuka H, et al. The expression of dentin sialophosphoprotein gene in bone. Journal of Dental Research. 2002;**81**:392-394

[76] Baba O, Qin C, Brunn JC, Jones JE, Wygant JN, BW MI, et al. Detection of dentin sialoprotein in rat periodontium. European Journal of Oral Sciences. 2004;**112**:163-170

[77] Kim JW, Hu JCC, Lee JI, et al. Mutational hot spot in the DSPP gene causing dentinogenesis imperfecta type II. Human Genetics. 2005;**116**:186-191

[78] Verdelis K, Ling Y, Sreenath T, Haruyama N, MacDougall M, Van der Meulen MC, et al. DSPP effects on in vivo bone mineralization. Bone. 2008;**43**:983-990

[79] Yamakoshi Y, Simmer JP. Structural features, processing mechanism and gene splice variants of dentin sialophosphoprotein. Japanese Dental Science Review. 2018;**54**:183-196

[80] Yamakoshi Y, Hu JC, Fukae M, Zhang H, Simmer JP. Dentin glycoprotein: The protein in the middle of the dentin sialophosphoprotein chimera. The Journal of Biological Chemistry. 2005;**280**:17472-17479

[81] Yamakoshi Y, Lu Y, Hu JC, Kim JW, Iwata T, Kobayashi K, et al. Porcine dentin sialophosphoprotein: Length polymorphisms, glycosylation, phosphorylation, and stability. The Journal of Biological Chemistry. 2008;**283**:14835-14844

[82] He G, Ramachandran A, Dahl T, George S, Schultz D, Cookson D, et al.

Phosphorylation of phosphophoryn is crucial for its function as a mediator of biomineralization. The Journal of Biological Chemistry. 2005;**280**:33109-33114

[83] Boskey AL, Maresca M, Doty S, Sabsay B, Veis A. Concentration dependent effects of dentin phosphophoryn in the regulation of in vitro hydroxyapatite formation and growth. Bone and Mineral. 1990;**11**:55-65

[84] Stetler-Stevenson WG, Veis A. Type I collagen shows a specific binding affinity for bovine dentin phosphophoryn. Calcified Tissue International. 1986;**38**:135-141

[85] Traub W, Jodaikin A, Arad T, Veis A, Sabsay B. Dentin phosphophoryn binding to collagen fibrils. Matrix. 1992;**12**:197-201

[86] Prasad M, Butler WT, Qin C. Dentin sialophosphoprotein in biomineralization. Connective Tissue Research. 2010;**51**:404-417

[87] Butler WT. Dentin matrix proteins. European Journal of Oral Sciences. 1998;**106**(Suppl 1):204-210

[88] Butler WT, Brunn JC, Qin C. Dentin extracellular matrix (ECM) proteins: Comparison to bone ECM and contribution to dynamics of dentinogenesis. Connective Tissue Research. 2003;**44**(Suppl 1):171-178

[89] Goldberg M, Kulkarni AB, Young M, Boskey A. Dentin: Structure, composition and mineralization. The role of dentin ECM in dentin formation and mineralization. Frontiers in Bioscience. 2011;**3**:711-735

[90] Moradian-Oldak J. Protein-mediated enamel mineralization. Frontiers in bioscience (Landmark edition). 2012;**17**:1996-2023

[91] Ye L, Le TQ, Zhu L, Butcher K, Schneider RA, Li W, et al. Amelogenins in human developing and mature dental pulp. Journal of Dental Research. 2006;**85**:814-818

[92] Delak K, Harcup C, Lakshminarayanan R, Sun Z, Fan YJ, Moradian-Oldak J, et al. The tooth enamel protein, porcine amelogenin, is an intrinsically disordered protein with an extended molecular configuration in the monomeric form. Biochemistry. 2009;**48**:2272-2281

[93] Paine ML, Luo W, Zhu DH, Bringas PJ, Snead ML. Functional domains for amelogenin revealed by compound genetic defects. Journal of Bone and Mineral Research. 2003;**18**:466-472

[94] Nanci A, Zalzal S, Lavoie P, Kunikata M, Chen W, Krebsbach PH, et al. Comparative immunochemical analyses of the developmental expression and distribution of ameloblastin and amelogenin in rat incisors. The Journal of Histochemistry and Cytochemistry. 1998;**46**:911-934

[95] Krebsbach PH, Lee SK, Matsuki Y, Kozak CA, Yamada KM, Yamada Y. Full length sequence, localization, and chromosomal mapping of ameloblastin: A novel tooth-specific gene. The Journal of Biological Chemistry. 1996;**271**:4431-4435

[96] Fukumoto S, Kiba T, Hall B, Iehara N, Nakamura T, Longenecker G, et al. Ameloblastin is a cell adhesion molecule required for maintaining the differentiation state of ameloblasts. The Journal of Cell Biology. 2004;**167**:973-983

[97] Vymetal J, Slaby I, Spahr A, Vondrasek J, Lyngstadaas SP. Bioinformatic analysis and molecular modelling of human ameloblastin suggest a two-domain intrinsically unstructured calcium-binding protein. European Journal of Oral Sciences. 2008;**116**:124-134

[98] Mazumder P, Prajapati S, Lokappa SB, Gallon V, Moradian-Oldak J. Analysis of co-assembly and co-localization of ameloblastin and amelogenin. Frontiers in Physiology. 2014;**5**:274

[99] Mazumder P, Prajapati S, Bapat R, Moradian-Oldak J. Amelogenin ameloblastin spatial interaction around maturing enamel rods. Journal of Dental Research. 2016;**95**(9):1042-1048

[100] Uchida T, Tanabe T, Fukae M, Shimizu M, Yamada M, Miake K, et al. Immunochemical and immunohistochemical studies, using antisera against porcine 25 kDa amelogenin, 89 kDa enamelin and the 13-17 kDa nonamelogenins, on immature enamel of the pig and rat. Histochemistry. 1991;**96**:129-138

[101] Uchida T, Murakami C, Dohi N, Wakida K, Satoda T, Takahashi O. Synthesis, secretion, degradation, and fate of ameloblastin during the matrix formation stage of the rat incisor as shown by immunocytochemistry and immunochemistry using region-specific antibodies. The Journal of Histochemistry and Cytochemistry. 1997;**45**:1329-1340

[102] Fang PA, Conway JF, Margolis HC, Simmer JP, Beniash E. Hierarchical self assembly of amelogenin and the regulation of biomineralization at the nanoscale. Proceedings of the National Academy of Sciences of the United States of America. 2011;**108**:14097-14102

[103] Wald T et al. Intrinsically disordered enamel matrix protein ameloblastin forms ribbon-like supramolecular structures via an N-terminal segment encoded by exon 5. The Journal of Biological Chemistry. 2013;**288**:22333-22345

[104] Wald T, Spoutil F, Osickova A, Prochazkova M, Benada O, Kasparek P, et al. Intrinsically disordered proteins drive enamel formation via an evolutionarily conserved self-assembly motif. Proceedings of the National Academy of Sciences. 2017;**114**:E1641-E1650

[105] MacDougall MJ, Javed A. Dentin and bone: Similar collagenous mineralized tissues. In: Bronner F, Farach-Carson M, Roach H, editors. Bone and Development. Topics in Bone Biology. Vol. 6. London: Springer; 2010

[106] Orsini G, Ruggeri A, Mazzoni A, Nato F, Manzoli L, Putignano A, et al. A review of the nature, role, and function of dentin non-collagenous proteins. Part 1: Proteoglycans and glycoproteins. Endodontic Topics. 2009;**21**:1-18

[107] Niu LN, Zhang W, Pashley DH, Breschi L, Mao J, Chen JH, et al. Biomimetic remineralization of dentin. Dental Materials. 2014;**30**:77-96

[108] Kinney JH, Marshall SJ, Marshall GW. The mechanical properties of human dentin: A critical review and re-evaluation of the dental literature. Critical Reviews in Oral Biology and Medicine. 2003;**14**:13-29

[109] Bosshardt DD, Selvig KA. Dental cementum: The dynamic tissue covering of the root. Periodontology 2000. 2000;**1997**(13):41-75

[110] Saygin NE, Giannobile WV, Somerman MJ. Molecular and cell biology of cementum. Periodontology 2000. 2000;**2000**(24):73-98

[111] Villarreal-Ramírez E, Moreno A, Mas-Oliva J, Chávez-Pacheco LJ, Narayanan AS, Gil-Chavarría I, et al. Characterization of recombinant human cementum protein 1 (hrCEMP1): Primary role in biomineralization. Biochemical and Biophysical Research Communications. 2009;**384**:49-54

[112] Stock SR. The mineral–collagen interface in bone. Calcified Tissue International. 2015;**97**:262-280

[113] Weiner S, Wagner HD. The material bone: Structure-mechanical function relations. Annual Review of Materials Science. 1998;**28**:271-298

[114] Akisaka T, Yoshida A. Ultrastructural analysis of apatite-degrading capability of extended invasive podosomes in resorbing osteoclasts. Micron. 2016;**88**:37-47

[115] Veillat V, Spuul P, Daubon T, Egaña I, Kramer I, Génot E. Podosomes: Multipurpose organelles? The International Journal of Biochemistry & Cell Biology. 2015;**65**:52-60

Chapter 5

Inorganic Polyphosphates Are Important for Cell Survival and Motility of Human Skin Keratinocytes and Play a Role in Wound Healing

Cynthia M. Simbulan-Rosenthal, Bonnie C. Carney, Anirudh Gaur, Manish Moghe, Elliott Crooke, Lauren T. Moffatt, Jeffrey W. Shupp and Dean S. Rosenthal

Abstract

Inorganic polyphosphate (polyP) is a simple ancient polymer of linear chains of orthophosphate residues linked by high energy phospho-anhydride bonds ubiquitously found in all organisms. Despite its structural simplicity, it plays diverse functional roles. polyP is involved in myriad of processes including serving as microbial phosphagens, buffer against alkalis, Ca^{2+} storage, metal-chelating agents, pathogen virulence, cell viability and proliferation, structural component and chemical chaperones, and in the microbial stress response. In mammalian cells, polyP has been implicated in blood coagulation, inflammation, bone differentiation, cell bioenergetics, signal transduction, Ca^{2+}-signaling, neuronal excitability, as a protein-stabilizing scaffold, and in wound healing, among others. This chapter will discuss (1) polyP metabolism and roles of polyP in prokaryotic and eukaryotic cells, (2) the contribution of polyP to survival, cell proliferation, and motility involved in wound healing in human skin keratinocytes, (3) the use of polyP-containing platelet-rich plasma (PRP) to promote wound healing in acute and chronic wounds, including burns, and (4) the use of polyP-containing PRP in excisional wound models to promote faster healing. While polyP shows promise as a therapeutic agent to accelerate healing for acute and chronic wounds, the molecular mechanisms as a potent modulator of the wound healing process remain to be elucidated.

Keywords: inorganic polyphosphate, wound healing, keratinocytes, platelet-rich plasma

1. Introduction

PolyP is a simple prebiotic molecule that varies in chain length between three and several thousand inorganic phosphates linked by phosphoanhydride bonds (**Figure 1A**). It is continuously synthesized from ATP or GTP and degraded by cellular enzymes in bacteria [1–5] and eukaryotes, yet its pleiotropic functions remain

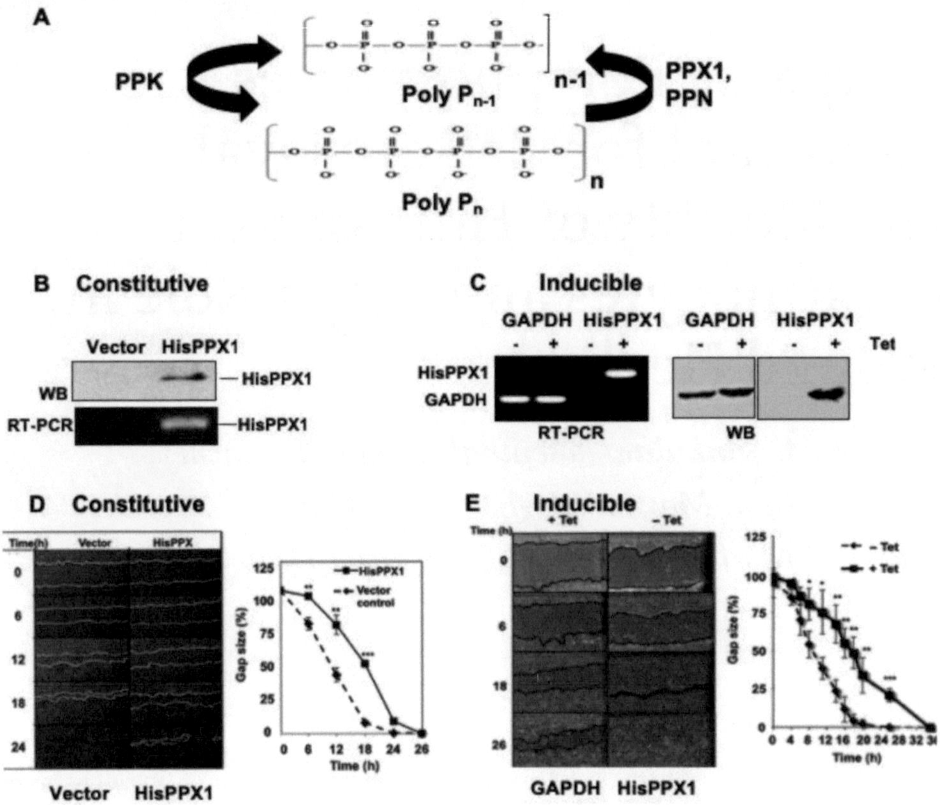

Figure 1.
(A) Schematic of linear structure of polyP, and its synthesis by polyphosphate kinase (PPK) and degradation by exopolyphosphatases (PPX1) or endopolyphosphatases (PPN). Constitutive (B and D) or Tet-inducible PPX1 (C and E) expression slows wound healing in cultured human keratinocytes (modified with permission from [33]).

to be clarified. In bacteria, but not mammalian cells, the enzymes that catalyze these activities have been identified [6–10]. polyP is synthesized following osmotic, oxidative, UVB, or other cellular stresses, and augments bacterial survival [11–16]; four roles have been proposed: an energy source, chelation of metal ions, storage of phosphate and response to cellular stresses. Necessary for bacterial survival, polyP is implicated in essential biological processes in prokaryotes including stress response, motility, biofilm formation [17–22], virulence, sporulation, and quorum-sensing [21–23]. Enzymes involved in polyP synthesis and degradation have been isolated and characterized in bacteria and other lower eukaryotes, and maintain tight control of polyP levels, as might be expected for a polymer controlling vital biological processes.

1.1 PolyP in prokaryotes

Drawing from the cellular ATP pool, bacterial polyphosphate kinase (PPK1) catalyzes the reversible transfer of the terminal γ-phosphate of ATP to polyP [24], whereas PPK2, expressed in other prokaryotes, transfers the terminal inorganic phosphate (Pi) from polyP to GDP to form GTP [4, 25, 26]. In contrast, the exopolyphosphatase PPX1 hydrolyzes polyP into phosphate monomers, thus maintaining phosphate homeostasis [5]. PPK levels and activity are tightly regulated, maintaining steady-state polyP concentrations in the bacterial cytosol at low micromolar levels, even in mutant strains deficient in PPX [27]. PolyP synthesis is upregulated during nutrient deprivation [28, 29], or during osmotic [28], acidic pH [30], oxidative [31], or heat [32] stresses, potentially depleting cellular ATP pools by converting millimolar levels of ATP to long polyP chains (>1000 Pi) [31]. polyP levels are measured using enzyme-based assays that employ ppk to generate ATP from ADP, using luciferase as a reporter [33]. In another assay, cells are labeled with $^{32}P_i$; polyP is isolated and hydrolyzed with PPX1, and thin-layer chromatography or phosphoimage analysis is performed [17]. Toluidine blue binding assays are effective for different chain-lengths but are relatively insensitive. Recently, a rapid and simple method has been described [34, 35]. ^{31}P-NMR Spectroscopy is an effective and accurate method for measuring polyP in intact cells [36]. Electron ionization mass spectrometry [37], cryoelectron tomography and spectroscopy imaging [38] have also been employed. Protein affinity labeling *in vivo* uses the affinity of a recombinant polyP-binding domain of *E. coli* PPX1 (PPXbd) [39], which we and other investigators have used to specifically inhibit the function of polyP.

Not surprisingly, *ppk* mutants are extremely sensitive to environmental stresses [13, 31, 32, 40, 41], and exhibit reduced motility, virulence, and biofilm production [42]. *ppk* gene expression is regulated by σ38, a transcriptional regulator for late stationary phase genes [43] and polyP, in turn, amplifies its own synthesis by inducing transcription of the gene encoding σ38 (RpoS) [29, 41, 42]. In response to oxidative or heat stress, polyP synthesis is regulated at a transcriptional and/or post-translational level, as PPK synthesis and levels are altered by antisense RNA that target *ppk* mRNA transcripts [44] as well as transient inactivation of PPX by stress-sensitive regulators that allow polyP levels to remain high until normal conditions are restored [45].

1.2 PolyP in eukaryotes

While polyP is found in eukaryotes from protists to mammalian cells [11], the mechanism of polyP synthesis remains largely unknown for most eukaryotic organisms [1, 45], except for *S. cerevisiae*, where the vacuolar transporter chaperone 4 (VTC4) synthesizes polyP from ATP and then transports the polymer into vacuoles [46, 47]. Vacuolar polyP maintains phosphate homeostasis by appropriating phosphate during growth in phosphate-rich conditions [46], and releases phosphate during the cell cycle to provide precursors for DNA replication [48].

There is no sequence or structural homology between the polyP-synthesizing enzymes PPK1, PPK2, or VTC4, and homologues have yet to be found in higher eukaryotes [1]. While phylogenetic analysis of the prokaryotic branch reveals no clear homologues of *E. coli* PPK in a large number of polyP-synthesizing species [49], a few enzymes have been identified that use polyP as phosphate donor in reactions that can be reversed in the presence of excess substrate *in vitro* [49, 50]. In the absence of a polyP-synthesizing enzyme, polyP may be synthesized by the mitochondrial proton-motive force [51], in a complex process involving intact mitochondrial membranes [52]. Decreased polyP production resulting from depolarization of the mitochondrial membrane [52, 53] suggests that this may be a spontaneous process that does not need catalysis. Alternatively, inositol phosphates have also been implicated in polyP metabolism, since polyP levels are diminished in cells lacking the enzyme that synthesizes highly phosphorylated inositols [54–57]. Similar to polyP synthesis, polyP-degrading enzymes, such as yeast PPX1, have been found in lower eukaryotes, while mammalian polyP-specific degradation enzymes are mostly uncharacterized. However, h-prune, which regulates cell migration, also acts as a exopolyphosphatase for short-chain polyP *in vitro* [58, 59].

Subcellular fractionation, immunofluorescent staining, and biochemical quantification reveal that polyP is localized to the nucleus, cell membrane, cytoplasm, and intracellular organelles in mammalian cells. It is specifically enriched in nucleoli, acidocalcisomes (organelles rich in protons, calcium (Ca^{2+}), and phosphorus), and mitochondria [11, 60–63]. In the brain, astrocytes secrete polyP, which is taken up by neurons, indicating both intra- and extracellular localization [64, 65]. Similar to vesicular packaging of ATP, astrocytes release polyP *via* exocytosis from vesicular nucleotide transporter (VNUT)-containing vesicles [65]. A putative G protein-coupled receptor in *D. discoideum* mediates cell surface binding of extracellular polyP, which as a signaling molecule, elicits differential effects on cell-substratum adhesion and cytoskeletal F-actin levels [66].

Eukaryotic polyP levels are in the 20–100 μM range (expressed as Pi concentration), with chain lengths ranging from 50 to 800 Pi residues (rat tissues) [67], averaging ~80 Pi residues in human platelets [68] to 200 Pi residues in yeast [67], compared to bacterial polyP, which can range up to thousands of Pi units long [53]; however, up to 130 mm medium-sized polyP chains are stored in dense granules in thrombocytes and mast cells [62, 68, 69]. Brain tissue exhibits among the highest polyP levels (~100 μm), which drop with age and neurodegenerative disease [11, 70–72], consistent with the role of polyP in stabilizing protein unfolding intermediates as amyloid-like precursors [32]. High levels of polyphosphate are also found in osteoblast matrix vesicles, the initial sites of bone mineral formation [73]. PolyP concentrations and chain lengths are dynamic, and depend on growth conditions of cells; for example in *Plasmodia*, polyP has an average chain length of 100 Pi, which is degraded to 10 Pi during sporulation [67].

Studies on the myriad roles of polyP in higher eukaryotes have recently gained momentum. PolyP is directly or indirectly involved in diverse cell processes, including control of cell bioenergetics, signal transduction, activation of the mitochondrial permeability transition pore (mPTP), Ca^{2+}-signaling [74, 75], and maintenance of the mitochondrial membrane potential [74]. Associated with mPTP [74] and voltage-gated channels, polyP regulates neuronal excitability [76] and astroglial signaling [64]. About 39% of intracellular polyP pools in astrocytes are in mitochondria [77], playing a role in bioenergetics [52, 77] and Ca^{2+}-handling [74, 78, 79]. As a signaling molecule, polyP released from astrocytes can mediate the physiological response to brain hypoxia [65].

In addition to its role as a gliotransmitter in the autonomic nervous system [64], the polymer also interacts with a variety of proteins, such as mammalian target of rapamycin (mTOR), fibroblast growth factor (FGF)-2, TRPM8, integrin β1, and glycosomal and ribosomal proteins and enzymes, consequently modulating cell survival and cell growth [80–86]. A fascinating finding of several recent studies is that polyP can covalently and non-enzymatically modify a small number of specific proteins in yeast [81] and humans containing lysine residues located in poly-acidic, serine, and lysine-rich (PASK) motifs, some of which are involved in ribosome biogenesis [87].

PolyP is involved in mTOR signaling, cell proliferation, and apoptosis [74, 80], and stimulates the mTOR pathway [80] at concentrations normally found in mammalian cells (0.15–1.5 mM); [11], suggesting a role for the polymer in mammalian cell proliferation. By promoting release of translation initiation factor eIF4E, mTOR stimulates initiation of translation, particularly proteins involved in cell growth and proliferation. PolyP also enhances the mitogenic activity of FGF-2 by promoting its binding to cell surface receptors [83]. PolyP appears to regulate apoptosis by inducing activation of caspase-3 in human plasma cells [88]. This polyanion also chelates metals, such as manganese and cadmium, blocking metal-induced cell damage [89, 90].

Other roles for polyP in mammalian cells include coagulation *via* activation of blood clotting factor XII [69], inflammation [91] as well as Ca^{2+} chelation for bone mineralization and osteogenic differentiation [92]. The polymer also contributes to pro-inflammatory responses upon release from mast cells [62]. Finally, serving as a stabilizing scaffold for protein-folding intermediates, polyP was recently shown to work as a protein-like chaperone protecting cells against stress-induced protein aggregation [93].

2. The contribution of polyP to cell survival, proliferation, and motility involved in wound healing in skin keratinocytes

Inorganic polyP shows promise in different phases of wound healing, including hemostasis and re-epithelialization, as polyP is a normal component of different cells that play a role in this process, including platelets, dermal fibroblasts, and keratinocytes. The use of polyP as a therapeutic for acute and chronic wounding has begun to garner interest, in part because of experiments elucidating its role in wound healing [33], as well as in hemostasis [69, 94–99]. We have recently shown a role for polyP in the response to UV survival, cell motility, and wound healing [33]. Addition of exogenous polyP increased the rate of wound healing in standard scratch wound assays *in vitro* [33].

Whereas candidates for mammalian polyP metabolism have been shown to exhibit additional enzymatic activities [59], more specific polyphosphatases have been identified in lower eukaryotes including yeast, trypanosomes, and *Dictyostelium*. We therefore used exopolyphosphatase derived from *S. cerevisiae* (ScPPX1) to target intracellular polyP in human skin keratinocytes, an obvious choice for UV resistance, motility, and wound healing. The functions of polyP in the response of keratinocytes to UVB or wounding was studied by expressing ScPPX1, which selectively breaks down endogenous inorganic polyP, and not phosphoproteins, DNA, RNA, or nucleotide mono-, di-, or triphosphates [6]. Cells depleted of intracellular polyP by ScPPX1 expression exhibited increased sensitivity to UVB *via* enhanced apoptosis, and impaired wound healing [33]. Human keratinocytes stably expressing constitutive HisPPX1 or tetracycline (Tet)-inducible HisPPX1 were used to deplete cells of endogenous polyP, and study its role in wound healing assays performed on confluent monolayers, mimicking cell re-epithelialization during wound healing *in vivo*. RT-PCR and immunoblot analysis confirmed PPX1 expression in stable HisPPX1-expressing cells or in the presence of Tet (**Figure 1B** and **C**). Scratch gaps demonstrate marked attenuation of wound healing following constitutive HisPPX1 or Tet-induced expression (**Figure 1D** and **E**).

Since keratinocyte proliferation and migration are crucial to re-epithelialization during wound healing, the contribution of polyP to cell growth and motility involved in wound healing was next determined. Vector control cells exhibited significantly higher rates of cell growth as well as BrdU incorporation into newly synthesized DNA in cells at the wound edge, compared with polyP-depleted HisPPX1-expressing cells (**Figure 2**). Further, real time monitoring and measurement of cell motility performed in an xCelligence impedance-based system revealed significant decreases in cell motility in polyP-depleted keratinocytes (**Figure 3**). These results demonstrate that polyP depletion by either constitutive or inducible expression of PPX1 retards the rate of wound healing in human skin keratinocytes, by decreasing cell proliferation and motility. To determine if the loss of endogenous polyP can be supplemented with exogenous extracellular polyP, ScPPX1-expressing cells were grown in the presence of different concentrations of polyP, or with polyP-rich platelet lysate (next section). Exogenously added polyP was found to accelerate wound healing in human keratinocytes in polyP dose-response experiments on confluent monolayers of keratinocytes subjected to scratch wound healing assays

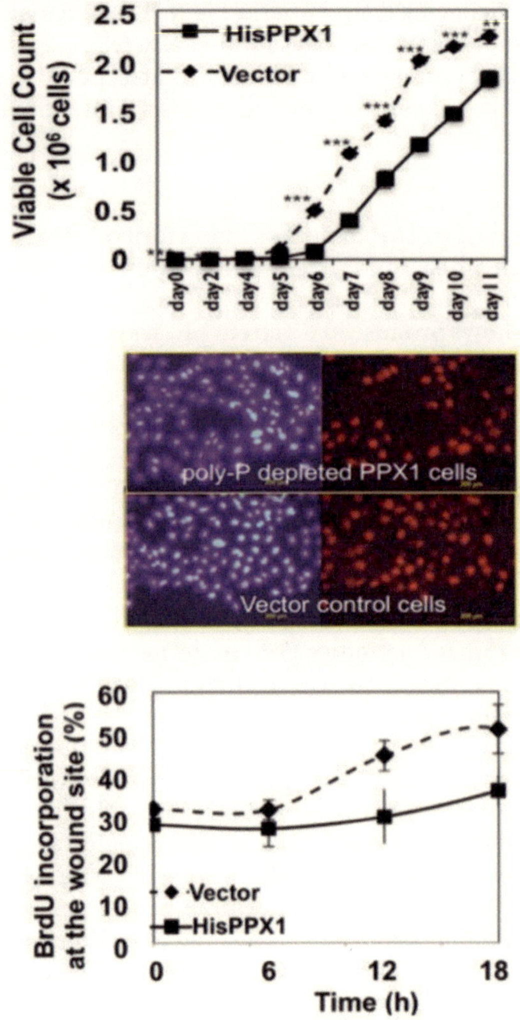

Figure 2.
Constitutive PPX1 *expression decreases growth rate of keratinocytes. Viable cell counts were performed over 11 days, and growth curves plotted for HisPPX1-expressing cells compared to vector cells (top). HisPPX1 and vector controls were subjected to scratch assays and proliferation was measured by in situ BrdU incorporation in cells at the wound edge (middle and bottom; modified with permission from [33]).*

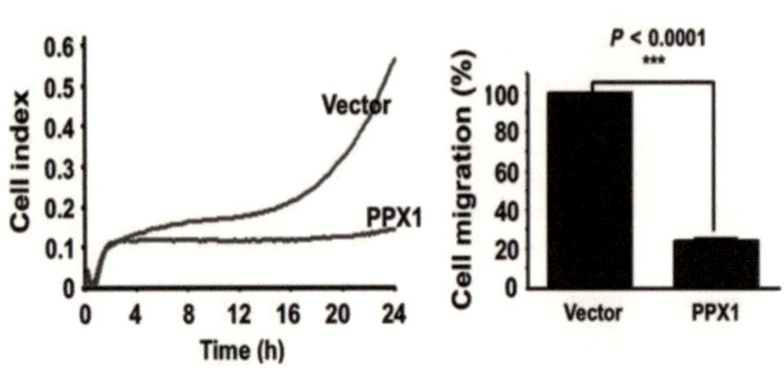

Figure 3.
Real-time monitoring and measurement of cell motility was performed in an xCelligence impedance-based system. Total number of cells attached to the bottom chamber were measured every 15 min over 24 hours, and are shown as technical duplicates, with assays repeated twice (left). Percentage of cell migration at the 24-hour time-point based on 100,000 cells plated at the top chamber (right; modified with permission from [33]).

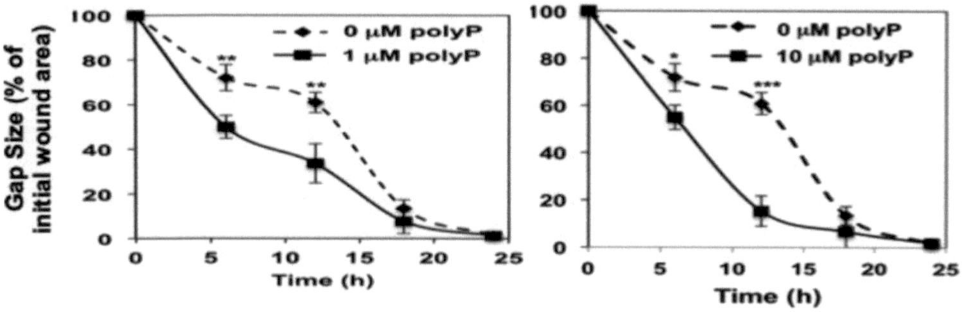

Figure 4.
PolyP dose-dependently accelerates wound healing in cultured human keratinocyte, using polyP at 1 µM (left) or 10 µM (right; modified with permission from [33]).

(**Figure 4**). Interestingly, both intracellular and extracellular polyP dose-dependently increased the rate of wound healing *in vitro*.

3. The use of polyP-containing PRP to promote wound healing in acute and chronic wounds, including burns

Wound healing is a highly coordinated process involving biochemical and physiological interplay of keratinocytes and fibroblasts to restore skin integrity. Platelets are components of blood responsible for blood clotting and wound healing. Although the importance of platelets in wound healing has been extensively studied, the bioactive substance playing a major role in skin re-epithelialization during wound healing is still unclear. PRP has been shown to support the survival and proliferation of human keratinocytes [100], and is currently used as therapeutic for both acute and chronic wounds (for review see [101]). Discovery of the release of growth factors triggered an interest in using PRP for wound healing, and platelet lysates have been examined as a replacement for using fetal bovine serum in cell cultures, which may contain contaminants such as prions, or elicit an unwanted immune response in patients. Platelet-rich lysates derived from platelets are a by-product of blood preparation, and are thus inexpensive. Most studies focused on lysates for mesenchymal stromal cell culture for cell therapy, in which platelets are activated by thrombin and $CaCl_2$, or by freeze-thaw.

In recent years, PRP has gained traction in many different specialties including in dermatology where it is used to treat acne [102], scarring [103], and alopecia [104, 105], in regenerative medicine where it is used to treat acute and chronic injuries to bone and cartilage [106, 107], in orthopedics and sports medicine where it is used to treat rotator cuff tears, osteoarthritis of the knee, hamstring injuries, and Achilles tendinopathy [108–110], in dentistry where it is used during tooth extractions, periodontal surgery, and dental implant surgery [111, 112], and more recently in wound healing to promote enhanced healing [113–118]. PRP is an autologous blood product generated from multiple rounds of centrifugation that serve to concentrate the number of platelets in plasma. PRP contains high concentrations of growth factors such as platelet-derived growth factor (PDGF), epidermal growth factor (EGF), and transforming growth factor beta (TGFβ) compared to plasma and whole blood [119, 120]. It also contains higher levels of pro- and anti-inflammatory cytokines that promote enhanced healing. Lastly, PRP is known to contain inorganic polyP which is continually synthesized from ATP or GTP, and is degraded by cellular enzymes in bacteria and eukaryotes.

The role of polyP secreted by platelets and present in PRP on cell proliferation and wound healing was investigated in human HaCaT keratinocytes co-transduced with either ScPPX1 or vector control, along with DsRed or GFP, respectively, as fluorescent markers in order to visualize and track cells that have reduced or normal levels of polyP. Cells stably expressing fluorescent-tagged DsRed-PPX1 or GFP-empty vector were incubated with platelet lysate (4%) supplemented with or without exogenous pure polyP (1 μM). In both vector-GFP control and polyP-depleted PPX1-DsRed cells treated with polyP, platelet lysate, or platelet lysate + polyP, cell growth curves revealed a significant increase in cell proliferation compared to untreated controls (data to be published elsewhere). PolyP quantification in platelet lysates using a micromolar polyP assay kit showed that a 4% platelet lysate contains ~8 μM polyP, which was within the range used for exogenously added polyP. This assay measures increase in fluorescence intensity (emission 550 nm, excitation 415 nm) of a PPD dye upon binding to polyP.

Cell migration/scratch assays were performed on PPX1-DsRed or vector-GFP control keratinocytes to assess the effects on wound healing and cell motility. Fluorescent pictures were taken at 10 min intervals for 36 hours using an EVOS FL time-lapse imaging system, and gap closure was quantified by Image J. In both GFP-vector cells and PPX1-expressing cells, the rate of wound closure in the scratch assays were significantly increased when cells were incubated either with platelet lysate alone, polyP alone, or both (data to be published elsewhere). These results together indicate that exogenous polyP, delivered either purified or from platelet-enriched plasma, can accelerate wound healing.

To assess whether the increased rate of wound healing is attributable to polyP in platelet lysates, specific polyP inhibitors (polyP-binding protein PPXbd or UHRA-9, a kind gift from Dr. James Morrissey) were utilized in wound healing assays. PPXbd, a recombinant polyP-binding domain of *E. coli* exopolyphosphatase, binds to platelet-derived polyP and blocks FXI activation, thrombin and fibrin generation, and consequently, inhibiting polyP procoagulant activity [99]. Interestingly, the enhanced rates of wound healing in vector control or polyP-depleted ScPPX-expressing cells induced by supplementation with exogenous extracellular polyP from pure polyP or in platelet lysates, was completely reversed by addition of the polyP inhibitors PPXbd or UHRA-9 (data to be published elsewhere). PolyP secreted by platelets and present in platelet lysate or PRP may therefore play an essential role in re-epithelialization during wound healing.

3.1 Chronic wounds

The acceleration of wound healing is of paramount importance in the setting of acute and chronic wounds, as well as burn wounds. Open chronic wounds are a significant cause of additional morbidity in patient populations that already have a plethora of comorbidities [121]. Significant improvements in complete healing were reported in chronic wounds treated with PRP compared to no topical treatments in a 2011 systematic review and meta-analysis on the use of PRP in acute and chronic wounds [101]. Another review of PubMed and Cochrane databases found significant benefit of PRP for diabetic chronic wounds, specifically in wounds unresponsive to standard of care treatment options [113]. A third systematic review of nine randomized controlled clinical trials (RCT) suggested that well-designed high-powered RCTs are needed to demonstrate increased wound healing with PRP treatment [122].

Treatment of 56 patients with diabetic foot ulcers with twice weekly applications of PRP resulted in complete healing in 86% of patients in the treated groups vs. only 68% in the control group [123]. Animal models using exosomes derived from PRP for full thickness skin wounds in a diabetic rat model also showed increased healing, as well as increased fibroblast proliferation and migration [124]. Platelet-rich fibrin also improved diabetic animal skin wound healing [125]. Overall, while there is no consensus on this treatment modality in chronic wounds, it is becoming widely used, and many trials seek to understand its potential beneficial effects. Improvements in open wound area have been shown in a number of animal and clinical studies.

3.2 Acute wounds

Meta-analysis of rodent and non-rodent studies using a systematic review conducted under preferred reported items for systematic review of interventions (PRISMA) guidelines indicated that the treatment of wounds with PRP resulted in reduction of open wound area [126]. In addition to its role in wound healing, PRP reduced complications such as wound infection, exudate (mass of cells and fluid that seeps out of a wound), drainage, and hematoma formation [101]. PRP and PRP with keratinocyte and fibroblast cells were shown to increase re-epithelialization at 7–14 days post-injury in mouse models, compared to non-treated controls [127]. In full thickness porcine wounds treated with the secreted proteins of PRP, wound re-epithelialization and collagen deposition were significantly increased in treated animals vs. saline controls [128]. Thus, PRP may improve wound healing in acute surgical wounds by secreting growth factors that support local microenvironments that promotes healing [129]. PRP's effectiveness has also been shown in bone grafting, cartilage regeneration, and non-cutaneous surgical procedures. The impact of PRP on normal and damaged (derived from chronic ulcers or irradiated) fibroblasts have been described [130]. In addition, despite the lack of reproducibility of platelet concentrations due to differences in manufacturer-specific protocols for PRP preparation and differences in treatment methodologies, PRP has been shown to affect fibroblast proliferation and migration in a number of *in vitro* studies. As with chronic wounds, it is unclear why some studies, but not others, show a beneficial effect of PRP treatment.

3.3 Burns

PRP has been used as a topical treatment to accelerate wound healing in burn wounds, however, like in chronic and acute wounds, its use is still debated due to conflicting results [114]. Some papers recommend its use [115, 117, 118, 131–135], while others have shown non-significant changes in outcomes after treatment with PRP, and advise caution in using it in a wide-spread manner [114, 136–138]. A review of PRP for burns concluded that PRP may be useful in regeneration of dermal structures, increasing graft-take, and increasing re-epithelialization, but recommended further research on characterization of the mechanisms by which PRP can improve burn wound healing, donor site healing, and scar outcomes [139].

The use of side-by-side treatment of a split thickness skin graft (STSG) donor site with standard treatment or with PRP showed complete re-epithelialization in the PRP-treated side at day 11 *vs.* day 13 for the control. Histological samples taken from these healing wounds, and by H and E staining revealed increased epidermal thickness in PRP-treated wounds, as well as a significant increase in the number of blood vessels. After platelet concentrate in conjunction with STSG was used for deep burns, monitoring of viscoelastic properties of the resultant scars over 12 months revealed that the skin's return to normal viscoelastic properties was accelerated in burns treated with PRP compared to controls [133]. Compared to historic institutional standard of care controls, treatment of deep partial thickness (DPT) burns using PRP applied with the autograft during skin grafting, pain scores, inflammation, pruritis (itchiness), cosmesis of the scar, and perfusion all showed improved outcomes [135].

Animal models of burn injury in rats treated with topical PRP or control showed that PRP treatment resulted in increased hydroxyproline, decreased inflammatory cells infiltration, but no difference in fibroblast collagen production or angiogenesis [134]. In a rat animal model of DPT burns, PRP was effective in increasing % wound closure, but showed little effectiveness in the full thickness injury group. PRP treatment resulted in increased neo-epidermal thickness at day 21, as well as decreases in CD31, 68, and 163, TGFβ1, MMP2, and MPO+ cells indicating an increased resolution of inflammation [118]. PRP injection in burn wound scars in a rat animal model of burn injury also showed pain-associated markers to be decreased with treatment [132]. While positive healing was observed in most models, in a recent study using a swine model of burn, tangential excision, and grafting with or without PRP, PRP showed similar effect on re-epithelialization and scarring in full thickness wounds compared to control wounds [136].

In a randomized clinical trial from 2018, 27 patients with DPT burns that did not get autografted were treated with lyophilized PRP powder, and showed a significant increased percent wound closure at 3 weeks in the treated group compared to control. Additionally, the infection rate in the PRP group was 26%, while 33% of control patients had postoperative infections [115]. PRP's concentrated secretion of growth factors may include basic fibroblast growth factor, epidermal growth factor, platelet-derived growth factor (PDGF), insulin-like growth factor, transforming growth factor β (TGFβ1), and vascular endothelial growth factor (VEGF) as probable mechanisms by which it can accelerate healing [139]. However, quantification of growth factors TGFβ1, PDGF-AA, and VEGF in a cohort of five burn patients compared to five healthy volunteers showed comparable levels of growth factors in the PRP from burn patients and health volunteers. Thus, there may be an additional factor in PRP that is possibly altered in burn patients (or patients with other pathologies such as diabetes or other conditions that would lead them to have surgical procedures yielding acute wounds) that may contribute to its success in treating some wounds and failure in others. Due to the effect burns have on the pathophysiology of blood coagulopathy [140–142] and capillary endotheliopathy [143, 144] after injury, it is reasonable to assume that platelets from burn patients may have differing levels of polyP. Associating this data with what is known about polyP, its ubiquitous presence in all prokaryotic and eukaryotic organisms, and its role in the response to cellular stress, it was hypothesized that polyP may contribute to wound healing.

4. PolyP and polyP-containing platelet rich plasma accelerates re-epithelialization *in vitro* and *in vivo*

A HaCaT keratinocyte polyP-depleted cell line and vector control was used in growth curves and scratch assays to evaluate polyP, platelet lysate, or combined treatment to accelerate wound healing *in vitro*. PolyP-containing PRP was also evaluated as a treatment in a splinted model of excisional wounding *in vivo*. Exogenous polyP was also spiked into PRP to assess its role. Treatment with the polyP-containing treatments increased cell growth and attenuated open wound area *in vitro* ($p < 0.001$). Addition of a polyP inhibitor abrogated these effects ($p < 0.0001$). PRP-treated wounds re-epithelialized faster compared to untreated wounds when analyzed at Days 3 and 5 ($n = 6$ wounds, $p < 0.05$). Re-epithelialization was further enhanced by exogenous polyP addition to PRP as evidenced by elongated epithelial tongues (**Figure 5**) in the low and high dose PRP + polyP treatment groups compared to PRP alone ($n = 8$ wounds, $p < 0.05$; data to be published elsewhere). Due to its autologous nature, PRP serves as a safe and efficacious option for accelerating wound healing, and may be enhanced by the exogenous addition of polyP.

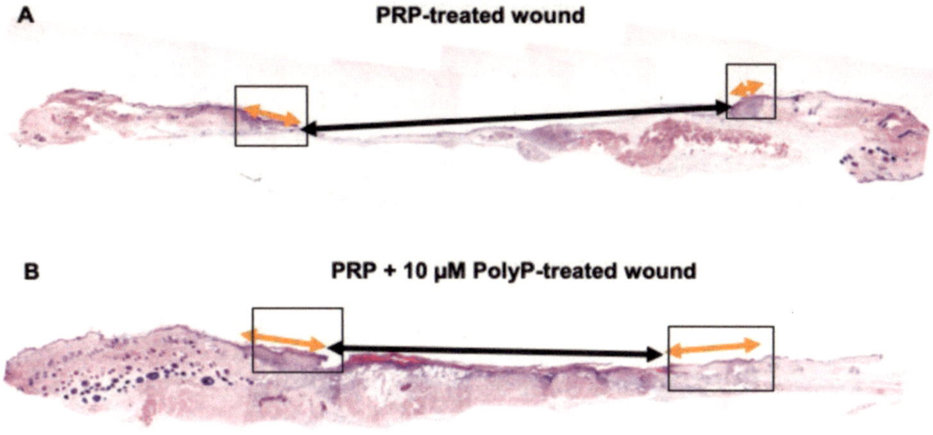

Figure 5.
Untreated or treated wounds were excised on day 5 and fixed in formalin, paraffin embedded, sectioned, and stained with H&E. Sections were imaged, and composited to create an image with areas of normal skin on both sides with the epithelial tongue (orange arrow) protruding from each side of normal skin. The epithelial gap is demarcated by the black arrow where no epithelium is present. Scale bar = 200 μm. Representative histology of untreated (A), PRP only treated (B), PRP + 10 μM polyP-treated wounds are shown. Epithelial tongue length was measured with Image J and quantified.

4.1 PRP-treated wounds heal faster than controls *in vivo*

PRP was generated based on a previously published protocol [145]. Briefly, whole blood was collected, and centrifuged to create platelet-poor plasma (PPP) and a pellet of platelets. PPP was then removed and the platelets were resuspended and activated with thrombin and calcium chloride ($CaCl_2$) to form a "biobandage"-like gel. Whole blood, packed RBCS, and PRP were stained with Wright and Giemsa stains to confirm PRP platelet concentration. A murine model of full thickness excisional wound healing was used where 6 mm punch biopsies were created on dorsal flanks of C57BL/6 mice [146, 147]. Wounds were splinted to encourage healing by re-epithelialization as opposed to contracture. They were subsequently treated with PRP, or no treatment was applied. Tegaderm dressing was applied and kept in place for 3 days. Pictures were taken at days 3, 4, 5, 6, or 7. By day 7, wounds were mostly closed.

In a second experiment, wounded mice were divided into four treatment groups: untreated, PRP only, PRP + 10 μM polyP (low dose) and PRP + 100 μM polyP (high dose). Doses of polyP were calculated by examining historical data from the literature on platelet levels, as well as polyP concentrations in platelets. The mean platelet count for C57BL/6 mice is $9.85 \pm 1.40 \times 10^{11}$ platelets/L, however, because there are no reliable reports of polyP concentrations per platelet in mice, human platelet counts and polyP concentrations were extrapolated for this experiment [148]. Human platelet concentrations range from 1.5 to 4.5×10^{11} platelets/L [149]. It is also known that platelets contain 0.74 ± 0.08 μmol polyP/1×10^{11} platelets [150]. Therefore, whole blood should contain between 1 and 3 μM polyP, and PRP, which contains at least 3-fold higher levels of platelets compared to whole blood, should contain 3–9 μM of polyP. We added ~10 μM in the low dose group, and a 10× concentration compared to the low dose group for the high dose group (100 μM). Lyophilized polyP was reconstituted in dH_2O to 1 M. PolyP took up 10% of the total treatment volume of PRP.

Treatments were applied on Day 0, and *Tegaderm*™ dressings stayed in place through day 3. On day 3 and 4, dressings were removed, and pictures were taken. On day 5, dressings were removed, pictures were taken, and wounds were excised with underlying fascia and were sewn into histological cassettes to retain wound orientation. These samples were then paraffin embedded and H&E stained. Sections were imaged, and epithelial tongue length was measured using Image J. Epithelial tongues were defined as new epithelium if there was no uninjured dermis underneath the epithelium.

Compared to whole blood, PRP contained a higher concentration of platelets (data to be published elsewhere). The splinted wound model was used to shift the healing towards re-epithelialization instead of the normal contraction observed in mice. PRP application was easily applied as a "bio-bandage" gel-like liquid (data to be published elsewhere). The 6 mm punch biopsies allowed for the creation of similar wound size between animal groups at day 0 ($0.33\ cm^2 \pm 0.13$ vs. $0.39 \pm 0.15\ cm^2$, p = n.s.; data to be published elsewhere). At days 3 (0.09 ± 0.06 vs. 0.23 ± 0.13) and 5 (0.12 ± 0.07 vs. 0.25 ± 0.12) PRP-treated wounds had significantly smaller open wound areas compared to control animals ($n = 6$, $p < 0.05$). At days 6 and 7, this difference leveled off.

4.2 Exogenous spiking of PRP with polyP further accelerates healing *in vivo*

To further investigate the potential role of polyP in PRP, wounds were treated with PRP or with PRP with low or high dose-spiked polyP. Untreated wounds were largely open by day 5, while PRP treated wounds were smaller and contained newly formed epithelium (data to be published elsewhere). Spiking with low or high dose polyP further stimulated epithelialization, and wounds were smaller with increasing doses. By histomorphometric analysis, epithelial tongues can be seen by H&E staining. In untreated wounds, these tongues are small and shallow. PRP treatment results in a more proliferative epithelium that is thicker and longer than untreated samples. Spiking with low and high dose polyP creates longer epithelial tongues). Epithelial tongue measurement by Image J shows a significant decrease in tongue length in untreated and PRP only treated vs. PRP + high dose polyP (737.38 ± 121.21 and 925.55 ± 214.17 vs. $1186.91 \pm 255.06\ \mu M$, $n = 8$, $p < 0.0001$, $p < 0.05$). PRP + high dose polyP-treated wounds also had significantly longer epithelial tongues compared to PRP + low dose polyP ($n = 8$, $p < 0.05$). PRP contains polyP at a concentration near 5 μM. The exogenous addition of polyP to the PRP promoted keratinocyte growth and proliferation, as is evidence by the increased epithelial tongue length with increasing doses of polyP administration.

5. Conclusion

PolyP plays key roles in essential biological processes in bacteria, and its increasing importance in eukaryotes is becoming apparent, including its participation in blood coagulation and wound healing. Recent advances in measurement and localization of polyP, along with our growing understanding of polyP metabolism and its interaction with specific proteins allows us to begin to analyze mechanisms responsible for cell-specific roles of polyP. Our ability to regulate polyP in eukaryotic cells opens possibilities for therapeutic intervention. Future work related to wound healing should be aimed at investigating the specific roles of intra- and extracellular polyP in keratinocytes, as well as the potential importance of polyP in other skin cells, including dermal fibroblasts, as these cells make up the majority of the skin. As polyP is also secreted by activated platelets and is important for normal blood clotting, the application of polyP-containing PRP as a biologic dressing may positively contribute to wound healing. We have completed two clinical trials using platelet-rich plasma for wound healing, a phenomenon that may be explained by the presence of polyP. PolyP levels and chain lengths should also be quantified in healthy and pathologic conditions in order to assess appropriate levels when treating acute or chronic wounds in the future.

Acknowledgements

Authors are grateful for the technical assistance of Sixian Song in the scratch gap assays, and Dr. James Morrissey (University of Michigan Medical School) for the generous gift of polyP, and its inhibitors PPXbd and UHRA-9, used in their studies. This work was partially supported by the NIH STTR grant 1R41ES026908 (to DSR), and the office of the Dean of Research, Georgetown University School of Medicine.

Conflict of interest

The authors have no conflicts of interest to declare.

Author details

Cynthia M. Simbulan-Rosenthal[1], Bonnie C. Carney[1,2], Anirudh Gaur[1], Manish Moghe[1], Elliott Crooke[1], Lauren T. Moffatt[1,2], Jeffrey W. Shupp[1,2,3,4] and Dean S. Rosenthal[1*]

1 Department of Biochemistry and Molecular and Cellular Biology, Georgetown University School of Medicine, Washington, DC, USA

2 Firefighters' Burn and Surgical Research Laboratory, MedStar Health Research Institute, Washington, DC, USA

3 The Burn Center, Department of Surgery, MedStar Washington Hospital Center, Washington, DC, USA

4 Department of Surgery, Georgetown University School of Medicine, Washington, DC, USA

*Address all correspondence to: rosenthd@georgetown.edu

References

[1] Rao NN, Gomez-Garcia MR, Kornberg A. Inorganic polyphosphate: Essential for growth and survival. Annual Review of Biochemistry. 2009;**78**:605-647. DOI: 10.1146/annurev.biochem.77.083007.093039

[2] Brown MR, Kornberg A. The long and short of it: Polyphosphate PPK and bacterial survival. Trends in Biochemical Sciences. 2008;**33**(6):284-290. DOI: 10.1016/j.tibs.2008.04.005

[3] Gomez-Garcia MR, Kornberg A. Formation of an actin-like filament concurrent with the enzymatic synthesis of inorganic polyphosphate. Proceedings of the National Academy of Sciences of the United States of America. 2004;**101**(45):15876-15880. DOI: 10.1073/pnas.0406923101

[4] Zhang H, Ishige K, Kornberg A. A polyphosphate kinase (PPK2) widely conserved in bacteria. Proceedings of the National Academy of Sciences of the United States of America. 2002;**99**(26):16678-16683. DOI: 10.1073/pnas.262655199

[5] Akiyama M, Crooke E, Kornberg A. An exopolyphosphatase of *Escherichia coli*. The enzyme and its ppx gene in a polyphosphate operon. The Journal of Biological Chemistry. 1993;**268**(1):633-639

[6] Wurst H, Kornberg A. A soluble exopolyphosphatase of *Saccharomyces cerevisiae*. Purification and characterization. The Journal of Biological Chemistry. 1994;**269**(15):10996-11001

[7] Wurst H, Shiba T, Kornberg A. The gene for a major exopolyphosphatase of *Saccharomyces cerevisiae*. Journal of Bacteriology. 1995;**177**(4):898-906

[8] Lichko LP, Kulakovskaya TV, Kulaev IS. Inorganic polyphosphate and exopolyphosphatase in the nuclei of *Saccharomyces cerevisiae*: Dependence on the growth phase and inactivation of the PPX1 and PPN1 genes. Yeast. 2006;**23**(10):735-740. DOI: 10.1002/yea.1391

[9] Sethuraman A, Rao NN, Kornberg A. The endopolyphosphatase gene: Essential in *Saccharomyces cerevisiae*. Proceedings of the National Academy of Sciences of the United States of America. 2001;**98**(15):8542-8547. DOI: 10.1073/pnas.151269398

[10] Luginbuehl E et al. The exopolyphosphatase TbrPPX1 of *Trypanosoma brucei*. BMC Microbiology. 2011;**11**:4. DOI: 10.1186/1471-2180-11-4

[11] Kumble KD, Kornberg A. Inorganic polyphosphate in mammalian cells and tissues. The Journal of Biological Chemistry. 1995;**270**(11):5818-5822

[12] Rao NN, Kornberg A. Inorganic polyphosphate regulates responses of *Escherichia coli* to nutritional stringencies, environmental stresses and survival in the stationary phase. Progress in Molecular and Subcellular Biology. 1999;**23**:183-195

[13] Crooke E et al. Genetically altered levels of inorganic polyphosphate in *Escherichia coli*. The Journal of Biological Chemistry. 1994;**269**(9):6290-6295

[14] Alcantara C et al. Accumulation of polyphosphate in *Lactobacillus* spp. and its involvement in stress resistance. Applied and Environmental Microbiology. 2014;**80**(5):1650-1659. DOI: 10.1128/AEM.03997-13

[15] Nikel PI et al. Accumulation of inorganic polyphosphate enables stress endurance and catalytic vigour in *Pseudomonas putida* KT2440. Microbial Cell Factories. 2013;**12**:50. DOI: 10.1186/1475-2859-12-50

[16] Singh R et al. Polyphosphate deficiency in *Mycobacterium tuberculosis* is associated with enhanced drug susceptibility and impaired growth in Guinea pigs. Journal of Bacteriology. 2013;**195**(12):2839-2851. DOI: 10.1128/JB.00038-13

[17] Rao NN, Liu S, Kornberg A. Inorganic polyphosphate in *Escherichia coli*: The phosphate regulon and the stringent response. Journal of Bacteriology. 1998;**180**(8):2186-2193

[18] Rao NN, Kornberg A. Inorganic polyphosphate supports resistance and survival of stationary-phase *Escherichia coli*. Journal of Bacteriology. 1996;**178**(5):1394-1400

[19] Rashid MH, Kornberg A. Inorganic polyphosphate is needed for swimming, swarming, and twitching motilities of *Pseudomonas aeruginosa*. Proceedings of the National Academy of Sciences of the United States of America. 2000;**97**(9):4885-4890. DOI: 10.1073/pnas.060030097

[20] Rashid MH, Rao NN, Kornberg A. Inorganic polyphosphate is required for motility of bacterial pathogens. Journal of Bacteriology. 2000;**182**(1):225-227

[21] Shi X, Rao NN, Kornberg A. Inorganic polyphosphate in *Bacillus cereus*: Motility, biofilm formation, and sporulation. Proceedings of the National Academy of Sciences of the United States of America. 2004;**101**(49):17061-17065. DOI: 10.1073/pnas.0407787101

[22] Rashid MH et al. Polyphosphate kinase is essential for biofilm development, quorum sensing, and virulence of *Pseudomonas aeruginosa*. Proceedings of the National Academy of Sciences of the United States of America. 2000;**97**(17):9636-9641. DOI: 10.1073/pnas.170283397

[23] Kim KS et al. Inorganic polyphosphate is essential for long-term survival and virulence factors in *Shigella* and *Salmonella* spp. Proceedings of the National Academy of Sciences of the United States of America. 2002;**99**(11):7675-7680. DOI: 10.1073/pnas.112210499

[24] Ahn K, Kornberg A. Polyphosphate kinase from *Escherichia coli*. Purification and demonstration of a phosphoenzyme intermediate. The Journal of Biological Chemistry. 1990;**265**(20):11734-11739

[25] Ishige K, Zhang H, Kornberg A. Polyphosphate kinase (PPK2), a potent, polyphosphate-driven generator of GTP. Proceedings of the National Academy of Sciences of the United States of America. 2002;**99**(26):16684-16688. DOI: 10.1073/pnas.262655299

[26] Nocek B et al. Polyphosphate-dependent synthesis of ATP and ADP by the family-2 polyphosphate kinases in bacteria. Proceedings of the National Academy of Sciences of the United States of America. 2008;**105**(46):17730-17735. DOI: 10.1073/pnas.0807563105

[27] Rudat AK et al. Mutations in *Escherichia coli* polyphosphate kinase that lead to dramatically increased in vivo polyphosphate levels. Journal of Bacteriology. 2018;**200**(6):pii: e00697-17. DOI: 10.1128/JB.00697-17

[28] Ault-Riche D et al. Novel assay reveals multiple pathways regulating stress-induced accumulations of inorganic polyphosphate in *Escherichia coli*. Journal of Bacteriology. 1998;**180**(7):1841-1847

[29] Shiba T et al. Inorganic polyphosphate and the induction of rpoS expression. Proceedings of the National Academy of Sciences of the United States of America. 1997;**94**(21):11210-11215

[30] Mullan A, Quinn JP, McGrath JW. Enhanced phosphate uptake and polyphosphate accumulation in

Burkholderia cepacia grown under low pH conditions. Microbial Ecology. 2002;**44**(1):69-77. DOI: 10.1007/s00248-002-3004-x

[31] Gray MJ et al. Polyphosphate is a primordial chaperone. Molecular Cell. 2014;**53**(5):689-699. DOI: 10.1016/j.molcel.2014.01.012

[32] Yoo NG et al. Polyphosphate stabilizes protein unfolding intermediates as soluble amyloid-like oligomers. Journal of Molecular Biology. 2018;**430**(21):4195-4208. DOI: 10.1016/j.jmb.2018.08.016

[33] Simbulan-Rosenthal CM et al. Inorganic polyphosphates are important for cell survival and motility of human skin keratinocytes. Experimental Dermatology. 2015;**24**(8):636-639. DOI: 10.1111/exd.12729

[34] Werner TP, Amrhein N, Freimoser FM. Novel method for the quantification of inorganic polyphosphate (iPoP) in *Saccharomyces cerevisiae* shows dependence of iPoP content on the growth phase. Archives of Microbiology. 2005;**184**(2):129-136. DOI: 10.1007/s00203-005-0031-2

[35] Freimoser FM et al. Systematic screening of polyphosphate (poly P) levels in yeast mutant cells reveals strong interdependence with primary metabolism. Genome Biology. 2006;**7**(11):R109. DOI: 10.1186/gb-2006-7-11-r109

[36] Zakrzewska J, Zizic M, Zivic M. The effect of anoxia on PolyP content of *Phycomyces blakesleeanus* mycelium studied by ^{31}P NMR spectroscopy. Annals of the New York Academy of Sciences. 2005;**1048**:482-486. DOI: 10.1196/annals.1342.073

[37] Choi BK, Hercules DM, Houalla M. Characterization of polyphosphates by electrospray mass spectrometry. Analytical Chemistry. 2000;**72**(20):5087-5091

[38] Comolli LR, Kundmann M, Downing KH. Characterization of intact subcellular bodies in whole bacteria by cryo-electron tomography and spectroscopic imaging. Journal of Microscopy. 2006;**223**(Pt 1):40-52. DOI: 10.1111/j.1365-2818.2006.01597.x

[39] Saito K et al. Direct labeling of polyphosphate at the ultrastructural level in *Saccharomyces cerevisiae* by using the affinity of the polyphosphate binding domain of *Escherichia coli* exopolyphosphatase. Applied and Environmental Microbiology. 2005;**71**(10):5692-5701. DOI: 10.1128/AEM.71.10.5692-5701.2005

[40] Groitl B et al. *Pseudomonas aeruginosa* defense systems against microbicidal oxidants. Molecular Microbiology. 2017;**106**(3):335-350. DOI: 10.1111/mmi.13768

[41] Dahl JU et al. The anti-inflammatory drug mesalamine targets bacterial polyphosphate accumulation. Nature Microbiology. 2017;**2**:16267. DOI: 10.1038/nmicrobiol.2016.267

[42] Fraley CD et al. A polyphosphate kinase 1 (ppk1) mutant of *Pseudomonas aeruginosa* exhibits multiple ultrastructural and functional defects. Proceedings of the National Academy of Sciences of the United States of America. 2007;**104**(9):3526-3531. DOI: 10.1073/pnas.0609733104

[43] Maciag A et al. In vitro transcription profiling of the sigmaS subunit of bacterial RNA polymerase: Re-definition of the sigmaS regulon and identification of sigmaS-specific promoter sequence elements. Nucleic Acids Research. 2011;**39**(13):5338-5355. DOI: 10.1093/nar/gkr129

[44] Silby MW, Nicoll JS, Levy SB. Regulation of polyphosphate kinase production by antisense RNA in *Pseudomonas fluorescens* Pf0-1. Applied and Environmental Microbiology.

2012;**78**(12):4533-4537. DOI: 10.1128/AEM.07836-11

[45] Kuroda A, Kornberg A. Polyphosphate kinase as a nucleoside diphosphate kinase in *Escherichia coli* and *Pseudomonas aeruginosa*. Proceedings of the National Academy of Sciences of the United States of America. 1997;**94**(2):439-442

[46] Ogawa N, DeRisi J, Brown PO. New components of a system for phosphate accumulation and polyphosphate metabolism in *Saccharomyces cerevisiae* revealed by genomic expression analysis. Molecular Biology of the Cell. 2000;**11**(12):4309-4321. DOI: 10.1091/mbc.11.12.4309

[47] Hothorn M et al. Catalytic core of a membrane-associated eukaryotic polyphosphate polymerase. Science. 2009;**324**(5926):513-516. DOI: 10.1126/science.1168120

[48] Bru S et al. Polyphosphate is involved in cell cycle progression and genomic stability in *Saccharomyces cerevisiae*. Molecular Microbiology. 2016;**101**(3):367-380. DOI: 10.1111/mmi.13396

[49] Wang L et al. Distribution patterns of polyphosphate metabolism pathway and its relationships with bacterial durability and virulence. Frontiers in Microbiology. 2018;**9**:782. DOI: 10.3389/fmicb.2018.00782

[50] Nahalka J, Patoprsty V. Enzymatic synthesis of sialylation substrates powered by a novel polyphosphate kinase (PPK3). Organic & Biomolecular Chemistry. 2009;**7**(9):1778-1780. DOI: 10.1039/b822549b

[51] Pestov NA, Kulakovskaya TV, Kulaev IS. Inorganic polyphosphate in mitochondria of *Saccharomyces cerevisiae* at phosphate limitation and phosphate excess. FEMS Yeast Research. 2004;**4**(6):643-648. DOI: 10.1016/j.femsyr.2003.12.008

[52] Pavlov E et al. Inorganic polyphosphate and energy metabolism in mammalian cells. The Journal of Biological Chemistry. 2010;**285**(13):9420-9428. DOI: 10.1074/jbc.M109.013011

[53] Kornberg A, Rao NN, Ault-Riche D. Inorganic polyphosphate: A molecule of many functions. Annual Review of Biochemistry. 1999;**68**:89-125. DOI: 10.1146/annurev.biochem.68.1.89

[54] Lonetti A et al. Identification of an evolutionarily conserved family of inorganic polyphosphate endopolyphosphatases. The Journal of Biological Chemistry. 2011;**286**(37):31966-31974. DOI: 10.1074/jbc.M111.266320

[55] Cordeiro CD, Saiardi A, Docampo R. The inositol pyrophosphate synthesis pathway in *Trypanosoma brucei* is linked to polyphosphate synthesis in acidocalcisomes. Molecular Microbiology. 2017;**106**(2):319-333. DOI: 10.1111/mmi.13766

[56] Ghosh S et al. Inositol hexakisphosphate kinase 1 maintains hemostasis in mice by regulating platelet polyphosphate levels. Blood. 2013;**122**(8):1478-1486. DOI: 10.1182/blood-2013-01-481549

[57] Hou Q et al. Inhibition of IP6K1 suppresses neutrophil-mediated pulmonary damage in bacterial pneumonia. Science Translational Medicine. 2018;**10**(435):pii: eaal4045. DOI: 10.1126/scitranslmed.aal4045

[58] Carotenuto M et al. H-prune through GSK-3beta interaction sustains canonical WNT/beta-catenin signaling enhancing cancer progression in NSCLC. Oncotarget. 2014;**5**(14):5736-5749. DOI: 10.18632/oncotarget.2169

[59] Tammenkoski M et al. Human metastasis regulator protein H-prune is a short-chain exopolyphosphatase.

Biochemistry. 2008;**47**(36):9707-9713. DOI: 10.1021/bi8010847

[60] Aschar-Sobbi R et al. High sensitivity, quantitative measurements of polyphosphate using a new DAPI-based approach. Journal of Fluorescence. 2008;**18**(5):859-866. DOI: 10.1007/s10895-008-0315-4

[61] Jimenez-Nunez MD et al. Myeloma cells contain high levels of inorganic polyphosphate which is associated with nucleolar transcription. Haematologica. 2012;**97**(8):1264-1271. DOI: 10.3324/haematol.2011.051409

[62] Moreno-Sanchez D et al. Polyphosphate is a novel pro-inflammatory regulator of mast cells and is located in acidocalcisomes. The Journal of Biological Chemistry. 2012;**287**(34):28435-28444. DOI: 10.1074/jbc.M112.385823

[63] Angelova PR et al. In situ investigation of mammalian inorganic polyphosphate localization using novel selective fluorescent probes JC-D7 and JC-D8. ACS Chemical Biology. 2014;**9**(9):2101-2110. DOI: 10.1021/cb5000696

[64] Holmstrom KM et al. Signalling properties of inorganic polyphosphate in the mammalian brain. Nature Communications. 2013;**4**:1362. DOI: 10.1038/ncomms2364

[65] Angelova PR et al. Signal transduction in astrocytes: Localization and release of inorganic polyphosphate. Glia. 2018;**66**(10):2126-2136. DOI: 10.1002/glia.23466

[66] Suess PM, Tang Y, Gomer RH. The putative G protein-coupled receptor GrlD mediates extracellular polyphosphate sensing in Dictyostelium discoideum. Molecular Biology of the Cell. 2019;**30**(9):1118-1128. DOI: 10.1091/mbc.E18-10-0686

[67] Chen KY. Study of polyphosphate metabolism in intact cells by 31-P nuclear magnetic resonance spectroscopy. Progress in Molecular and Subcellular Biology. 1999;**23**:253-273

[68] Ruiz FA et al. Human platelet dense granules contain polyphosphate and are similar to acidocalcisomes of bacteria and unicellular eukaryotes. The Journal of Biological Chemistry. 2004;**279**(43):44250-44257. DOI: 10.1074/jbc.M406261200

[69] Muller F et al. Platelet polyphosphates are proinflammatory and procoagulant mediators in vivo. Cell. 2009;**139**(6):1143-1156. DOI: 10.1016/j.cell.2009.11.001

[70] Gabel NW, Thomas V. Evidence for the occurrence and distribution of inorganic polyphosphates in vertebrate tissues. Journal of Neurochemistry. 1971;**18**(7):1229-1242

[71] Cremers CM et al. Polyphosphate: A conserved modifier of amyloidogenic processes. Molecular Cell. 2016;**63**(5):768-780. DOI: 10.1016/j.molcel.2016.07.016

[72] Lorenz B et al. Changes in metabolism of inorganic polyphosphate in rat tissues and human cells during development and apoptosis. Biochimica et Biophysica Acta. 1997;**1335**(1-2):51-60. DOI: 10.1016/S0304-4165(96)00121-3

[73] Li L et al. Long-chain polyphosphate in osteoblast matrix vesicles: Enrichment and inhibition of mineralization. Biochimica et Biophysica Acta: General Subjects. 2019;**1863**(1):199-209. DOI: 10.1016/j.bbagen.2018.10.003

[74] Abramov AY et al. Targeted polyphosphatase expression alters mitochondrial metabolism and inhibits calcium-dependent cell death. Proceedings of the National Academy of Sciences of the United States of America. 2007;**104**(46):18091-18096. DOI: 10.1073/pnas.0708959104

[75] Seidlmayer LK et al. Inorganic polyphosphate is a potent activator of the mitochondrial permeability transition pore in cardiac myocytes. The Journal of General Physiology. 2012;**139**(5):321-331. DOI: 10.1085/jgp.201210788

[76] Stotz SC et al. Inorganic polyphosphate regulates neuronal excitability through modulation of voltage-gated channels. Molecular Brain. 2014;**7**(1):42. DOI: 10.1186/1756-6606-7-42

[77] Angelova PR et al. Role of inorganic polyphosphate in mammalian cells: From signal transduction and mitochondrial metabolism to cell death. Biochemical Society Transactions. 2016;**44**(1):40-45. DOI: 10.1042/BST20150223

[78] Baev AY, Negoda A, Abramov AY. Modulation of mitochondrial ion transport by inorganic polyphosphate: Essential role in mitochondrial permeability transition pore. Journal of Bioenergetics and Biomembranes. 2017;**49**(1):49-55. DOI: 10.1007/s10863-016-9650-3

[79] Elustondo PA et al. Mitochondrial permeability transition pore induction is linked to formation of the complex of ATPase C-subunit, polyhydroxybutyrate and inorganic polyphosphate. Cell Death Discovery. 2016;**2**:16070. DOI: 10.1038/cddiscovery.2016.70

[80] Wang L et al. Inorganic polyphosphate stimulates mammalian TOR, a kinase involved in the proliferation of mammary cancer cells. Proceedings of the National Academy of Sciences of the United States of America. 2003;**100**(20):11249-11254. DOI: 10.1073/pnas.1534805100

[81] Azevedo C, Livermore T, Saiardi A. Protein polyphosphorylation of lysine residues by inorganic polyphosphate. Molecular Cell. 2015;**58**(1):71-82. DOI: 10.1016/j.molcel.2015.02.010

[82] Zakharian E et al. Inorganic polyphosphate modulates TRPM8 channels. PLoS ONE. 2009;**4**(4):e5404. DOI: 10.1371/journal.pone.0005404

[83] Shiba T et al. Modulation of mitogenic activity of fibroblast growth factors by inorganic polyphosphate. The Journal of Biological Chemistry. 2003;**278**(29):26788-26792. DOI: 10.1074/jbc.M303468200

[84] Segawa S et al. Probiotic-derived polyphosphate enhances the epithelial barrier function and maintains intestinal homeostasis through integrin-p38 MAPK pathway. PLoS ONE. 2011;**6**(8):e23278. DOI: 10.1371/journal.pone.0023278

[85] Negreiros RS et al. Inorganic polyphosphate interacts with nucleolar and glycosomal proteins in trypanosomatids. Molecular Microbiology. 2018;**110**(6):973-994. DOI: 10.1111/mmi.14131

[86] Azevedo C et al. Screening a protein Array with synthetic biotinylated inorganic polyphosphate to define the human PolyP-ome. ACS Chemical Biology. 2018;**13**(8):1958-1963. DOI: 10.1021/acschembio.8b00357

[87] Bentley-DeSousa A et al. A screen for candidate targets of lysine polyphosphorylation uncovers a conserved network implicated in ribosome biogenesis. Cell Reports. 2018;**22**(13):3427-3439. DOI: 10.1016/j.celrep.2018.02.104

[88] Hernandez-Ruiz L et al. Inorganic polyphosphate and specific induction of apoptosis in human plasma cells. Haematologica. 2006;**91**(9):1180-1186

[89] Trilisenko L, Kulakovskaya E, Kulakovskaya T. The cadmium tolerance in *Saccharomyces cerevisiae* depends on inorganic polyphosphate. Journal of

Basic Microbiology. 2017;**57**(11):982-986. DOI: 10.1002/jobm.201700257

[90] Andreeva N et al. Adaptation of *Saccharomyces cerevisiae* to toxic manganese concentration triggers changes in inorganic polyphosphates. FEMS Yeast Research. 2013;**13**(5):463-470. DOI: 10.1111/1567-1364.12049

[91] Dinarvand P et al. Polyphosphate amplifies proinflammatory responses of nuclear proteins through interaction with receptor for advanced glycation end products and P2Y1 purinergic receptor. Blood. 2014;**123**(6):935-945. DOI: 10.1182/blood-2013-09-529602

[92] Tsutsumi K et al. Morphogenetic study on the maturation of osteoblastic cell as induced by inorganic polyphosphate. PLoS ONE. 2014;**9**(2):e86834. DOI: 10.1371/journal.pone.0086834

[93] Xie L, Jakob U. Inorganic polyphosphate, a multifunctional polyanionic protein scaffold. The Journal of Biological Chemistry. 2019;**294**(6):2180-2190. DOI: 10.1074/jbc.REV118.002808

[94] Choi SH, Smith SA, Morrissey JH. Polyphosphate is a cofactor for the activation of factor XI by thrombin. Blood. 2011;**118**(26):6963-6970. DOI: 10.1182/blood-2011-07-368811

[95] Smith SA et al. Inhibition of polyphosphate as a novel strategy for preventing thrombosis and inflammation. Blood. 2012;**120**(26):5103-5110. DOI: 10.1182/blood-2012-07-444935

[96] Smith SA et al. Polyphosphate exerts differential effects on blood clotting, depending on polymer size. Blood. 2010;**116**(20):4353-4359. DOI: 10.1182/blood-2010-01-266791

[97] Smith SA, Morrissey JH. Polyphosphate: A new player in the field of hemostasis. Current Opinion in Hematology. 2014;**21**(5):388-394. DOI: 10.1097/MOH.0000000000000069

[98] Szymusiak M et al. Colloidal confinement of polyphosphate on gold nanoparticles robustly activates the contact pathway of blood coagulation. Bioconjugate Chemistry. 2016;**27**(1):102-109. DOI: 10.1021/acs.bioconjchem.5b00524

[99] Zhu S et al. FXIa and platelet polyphosphate as therapeutic targets during human blood clotting on collagen/tissue factor surfaces under flow. Blood. 2015;**126**(12):1494-1502. DOI: 10.1182/blood-2015-04-641472

[100] Baik SY et al. Effects of platelet lysate preparations on the proliferation of HaCaT cells. Annals of Laboratory Medicine. 2014;**34**(1):43-50. DOI: 10.3343/alm.2014.34.1.43

[101] Carter MJ, Fylling CP, Parnell LK. Use of platelet rich plasma gel on wound healing: A systematic review and meta-analysis. Eplasty. 2011;**11**:e38

[102] Porwal S, Chahar YS, Singh PK. A comparative study of combined dermaroller and platelet-rich plasma versus dermaroller alone in acne scars and assessment of quality of life before and after treatment. Indian Journal of Dermatology. 2018;**63**(5):403-408. DOI: 10.4103/ijd.IJD_118_17

[103] Alser OH, Goutos I. The evidence behind the use of platelet-rich plasma (PRP) in scar management: A literature review. Scars, Burns & Healing. 2018;**4**:2059513118808773. DOI: 10.1177/2059513118808773

[104] Garg S, Manchanda S. Platelet-rich plasma-an 'Elixir' for treatment of alopecia: Personal experience on 117 patients with review of literature. Stem Cell Investigation. 2017;**4**:64. DOI: 10.21037/sci.2017.06.07

[105] Alves R, Grimalt R. Double-blind, placebo-controlled pilot study

on the use of platelet-rich plasma in women with female androgenetic alopecia. Dermatologic Surgery. 2018;**44**(1):132-133. DOI: 10.1097/DSS.0000000000001197

[106] Liou JJ et al. Effect of platelet-rich plasma on chondrogenic differentiation of adipose- and bone marrow-derived mesenchymal stem cells. Tissue Engineering. Part A. 2018;**24**(19-20):1432-1443. DOI: 10.1089/ten.tea.2018.0065

[107] Xie X, Zhang C, Tuan RS. Biology of platelet-rich plasma and its clinical application in cartilage repair. Arthritis Research & Therapy. 2014;**16**(1):204. DOI: 10.1186/ar4493

[108] Chiavaras MM et al. IMpact of Platelet Rich plasma OVer alternative therapies in patients with lateral epicondylitis (IMPROVE): Protocol for a multicenter randomized controlled study: A multicenter, randomized trial comparing autologous platelet-rich plasma, autologous whole blood, dry needle tendon fenestration, and physical therapy exercises alone on pain and quality of life in patients with lateral epicondylitis. Academic Radiology. 2014;**21**(9):1144-1155. DOI: 10.1016/j.acra.2014.05.003

[109] Hussain N, Johal H, Bhandari M. An evidence-based evaluation on the use of platelet rich plasma in orthopedics: A review of the literature. SICOT J. 2017;**3**:57. DOI: 10.1051/sicotj/2017036

[110] Bansal H et al. Intra-articular injection in the knee of adipose derived stromal cells (stromal vascular fraction) and platelet rich plasma for osteoarthritis. Journal of Translational Medicine. 2017;**15**(1):141. DOI: 10.1186/s12967-017-1242-4

[111] Huang Y et al. Platelet-rich plasma for regeneration of neural feedback pathways around dental implants: A concise review and outlook on future possibilities. International Journal of Oral Science. 2017;**9**(1):1-9. DOI: 10.1038/ijos.2017.1

[112] Agrawal AA. Evolution, current status and advances in application of platelet concentrate in periodontics and implantology. World Journal of Clinical Cases. 2017;**5**(5):159-171. DOI: 10.12998/wjcc.v5.i5.159

[113] Picard F et al. The growing evidence for the use of platelet-rich plasma on diabetic chronic wounds: A review and a proposal for a new standard care. Wound Repair and Regeneration. 2015;**23**(5):638-643. DOI: 10.1111/wrr.12317

[114] Picard F et al. Should we use platelet-rich plasma as an adjunct therapy to treat "acute wounds," "burns," and "laser therapies": A review and a proposal of a quality criteria checklist for further studies. Wound Repair and Regeneration. 2015;**23**(2):163-170. DOI: 10.1111/wrr.12266

[115] Yeung CY et al. Efficacy of lyophilised platelet-rich plasma powder on healing rate in patients with deep second degree burn injury: A prospective double-blind randomized clinical trial. Annals of Plastic Surgery. 2018;**80**(2S Suppl 1):S66-S69. DOI: 10.1097/SAP.0000000000001328

[116] Hara T et al. Platelet-rich plasma stimulates human dermal fibroblast proliferation via a Ras-dependent extracellular signal-regulated kinase 1/2 pathway. Journal of Artificial Organs. 2016;**19**(4):372-377. DOI: 10.1007/s10047-016-0913-x

[117] Kakudo N et al. Platelet-rich plasma promotes epithelialization and angiogenesis in a splitthickness skin graft donor site. Medical Molecular Morphology. 2011;**44**(4):233-236. DOI: 10.1007/s00795-010-0532-1

[118] Venter NG et al. Use of platelet-rich plasma in deep second- and third-degree burns. Burns. 2016;**42**(4):807-814. DOI: 10.1016/j.burns.2016.01.002

[119] Alves R, Grimalt R. A review of platelet-rich plasma: History, biology, mechanism of action, and classification. Skin Appendage Disorders. 2018;**4**(1):18-24. DOI: 10.1159/000477353

[120] Arnoczky SP, Sheibani-Rad S. The basic science of platelet-rich plasma (PRP): What clinicians need to know. Sports Medicine and Arthroscopy Review. 2013;**21**(4):180-185. DOI: 10.1097/JSA.0b013e3182999712

[121] Martinengo L et al. Prevalence of chronic wounds in the general population: Systematic review and meta-analysis of observational studies. Annals of Epidemiology. 2019;**29**:8-15. DOI: 10.1016/j.annepidem.2018.10.005

[122] Brick N. Autologous platelet-rich plasma for treating chronic wounds. The American Journal of Nursing. 2013;**113**(8):54. DOI: 10.1097/01.NAJ.0000432965.18634.85

[123] Ahmed M et al. Platelet-rich plasma for the treatment of clean diabetic foot ulcers. Annals of Vascular Surgery. 2017;**38**:206-211. DOI: 10.1016/j.avsg.2016.04.023

[124] Guo SC et al. Exosomes derived from platelet-rich plasma promote the re-epithelization of chronic cutaneous wounds via activation of YAP in a diabetic rat model. Theranostics. 2017;7(1):81-96. DOI: 10.7150/thno.16803

[125] Ding Y et al. Platelet-rich fibrin accelerates skin wound healing in diabetic mice. Annals of Plastic Surgery. 2017;**79**(3):e15-e19. DOI: 10.1097/SAP.0000000000001091

[126] Tambella AM et al. Platelet-rich plasma to treat experimentally-induced skin wounds in animals: A systematic review and meta-analysis. PLoS ONE. 2018;**13**(1):e0191093. DOI: 10.1371/journal.pone.0191093

[127] Law JX et al. Platelet-rich plasma with keratinocytes and fibroblasts enhance healing of full-thickness wounds. Journal of Tissue Viability. 2017;**26**(3):208-215. DOI: 10.1016/j.jtv.2017.05.003

[128] Long DW et al. Controlled delivery of platelet-derived proteins enhances porcine wound healing. Journal of Controlled Release. 2017;**253**:73-81. DOI: 10.1016/j.jconrel.2017.03.021

[129] Fernandez-Moure JS et al. Platelet-rich plasma: A biomimetic approach to enhancement of surgical wound healing. The Journal of Surgical Research. 2017;**207**:33-44. DOI: 10.1016/j.jss.2016.08.063

[130] Devereaux J et al. Effects of platelet-rich plasma and platelet-poor plasma on human dermal fibroblasts. Maturitas. 2018;**117**:34-44. DOI: 10.1016/j.maturitas.2018.09.001

[131] Maciel FB et al. Scanning electron microscopy and microbiological evaluation of equine burn wound repair after platelet-rich plasma gel treatment. Burns. 2012;**38**(7):1058-1065. DOI: 10.1016/j.burns.2012.02.029

[132] Huang SH et al. Platelet-rich plasma injection in burn scar areas alleviates neuropathic scar pain. International Journal of Medical Sciences. 2018;**15**(3):238-247. DOI: 10.7150/ijms.22563

[133] Klosova H et al. Objective evaluation of the effect of autologous platelet concentrate on post-operative scarring in deep burns. Burns. 2013;**39**(6):1263-1276. DOI: 10.1016/j.burns.2013.01.020

[134] Ozcelik U et al. Effect of topical platelet-rich plasma on burn healing after partial-thickness burn

injury. Medical Science Monitor. 2016;**22**:1903-1909

[135] Prochazka V et al. Addition of platelet concentrate to dermo-epidermal skin graft in deep burn trauma reduces scarring and need for revision surgeries. Biomedical Papers of the Medical Faculty of the University Palacky, Olomouc, Czech Republic. 2014;**158**(2):242-258. DOI: 10.5507/bp.2013.070

[136] Singer AJ et al. The effects of platelet rich plasma on healing of full thickness burns in swine. Burns. 2018;**44**(6):1543-1550. DOI: 10.1016/j.burns.2018.04.021

[137] Marck RE, Middelkoop E, Breederveld RS. Considerations on the use of platelet-rich plasma, specifically for burn treatment. Journal of Burn Care & Research. 2014;**35**(3):219-227. DOI: 10.1097/BCR.0b013e31829b334e

[138] Marck RE et al. The application of platelet-rich plasma in the treatment of deep dermal burns: A randomized, double-blind, intra-patient controlled study. Wound Repair and Regeneration. 2016;**24**(4):712-720. DOI: 10.1111/wrr.12443

[139] Pallua N, Wolter T, Markowicz M. Platelet-rich plasma in burns. Burns. 2010;**36**(1):4-8. DOI: 10.1016/j.burns.2009.05.002

[140] Tejiram S et al. In-depth analysis of clotting dynamics in burn patients. The Journal of Surgical Research. 2016;**202**(2):341-351. DOI: 10.1016/j.jss.2016.01.006

[141] Shupp JW et al. Analysis of factor XIa, factor IXa and tissue factor activity in burn patients. Burns. 2018;**44**(2):436-444. DOI: 10.1016/j.burns.2017.08.003

[142] Glas GJ, Levi M, Schultz MJ. Coagulopathy and its management in patients with severe burns. Journal of Thrombosis and Haemostasis. 2016;**14**(5):865-874. DOI: 10.1111/jth.13283

[143] Luker JN et al. Shedding of the endothelial glycocalyx is quantitatively proportional to burn injury severity. Annals of Burns and Fire Disasters. 2018;**31**(1):17-22

[144] Vigiola Cruz M et al. Plasma ameliorates endothelial dysfunction in burn injury. The Journal of Surgical Research. 2019;**233**:459-466. DOI: 10.1016/j.jss.2018.08.027

[145] Messora MR, Nagata MJH, Furlaneto FAC, Dornelles RCM, Bomfim SRM, Deliberador TM, et al. A standardized research protocol for plateletrich plasma (PRP) preparation in rats. Revista Brasileira De Saude Ocupacional. 2011;**8**(3):299-304

[146] Galiano RD et al. Quantitative and reproducible murine model of excisional wound healing. Wound Repair and Regeneration. 2004;**12**(4):485-492. DOI: 10.1111/j.1067-1927.2004.12404.x

[147] Wang X et al. The mouse excisional wound splinting model, including applications for stem cell transplantation. Nature Protocols. 2013;**8**(2):302-309. DOI: 10.1038/nprot.2013.002

[148] Barrios M et al. Comparative hemostatic parameters in BALB/c, C57BL/6 and C3H/He mice. Thrombosis Research. 2009;**124**(3):338-343. DOI: 10.1016/j.thromres.2008.11.001

[149] Giacomini A et al. Platelet count and parameters determined by the Bayer ADVIA 120 in reference subjects and patients. Clinical and Laboratory Haematology. 2001;**23**(3):181-186

[150] Morrissey JH, Choi SH, Smith SA. Polyphosphate: An ancient molecule that links platelets, coagulation, and inflammation. Blood. 2012;**119**(25):5972-5979. DOI: 10.1182/blood-2012-03-306605

Chapter 6

Phosphonates: Their Natural Occurrence and Physiological Role

Paweł Kafarski

Abstract

The first natural compound containing carbon-to-phosphorus bond—ciliatine was discovered 60 years ago, and for four decades, phosphonates were considered simply as a biological curiosity. Finding the importance of these compounds in biogeochemical phosphorus cycling, their role in methane production, as well as discovery of numerous phosphonates and phosphonopeptides of promising antibacterial and antifungal activities has stimulated the development of studies on this class of compounds, especially on their metabolism and biochemistry. These studies are driven by the use of ^{31}P NMR and by a clever combination of genomics and innovative chemistry by using the method of selective labeling of metabolites. These studies revealed unusual and interesting chemistry of these compounds.

Keywords: C—P bond, phosphonates, ciliatine, phosphonopeptides, mimetics, antibiotics, ^{31}P NMR, genome mining

1. Introduction

Phosphonates are organophosphorus compounds characterized by a stable carbon-to-phosphorus (C—P) bond, which usually resists biochemical, thermal, and photochemical decomposition. The first phosphonate (compound **1**, **Figure 1**), being an analog of β-alanine and taurine, was isolated in 1959 from ciliated protozoa in the rumen of sheep [1]. That was the cause why its discoverers—M. Horiguchi

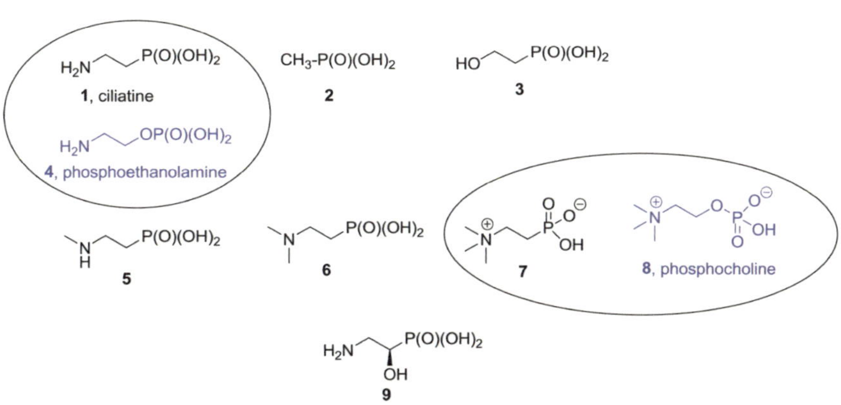

Figure 1.
Ciliatine (2-aminoethylphosphonic acids) and its derivatives found in lipids, glycans, glycoproteins, and bile acids.

and M. Kandatsu, named it ciliatine. This amino acid was then considered as a possible marker of the content of protozoa in sheep rumen, which appeared further to be misleading. For many years, natural compounds containing the C—P bond had been considered as curiosity being only scarcely studied. This is not the case in science currently because of their involvement in the global phosphorus cycle and in oceanic methane production. Some aspects of their occurrence, environmental role, biochemistry, and biological functions have been reviewed [2–5]. This chapter will concentrate on discussion of chemical diversity of the naturally occurring phosphonates and on the indication of open problems, which have not yet been solved.

2. Occurrence of carbon-to-phosphorus bond

The discovery of ciliatine stimulated intensive studies on the distribution of phosphonates in nature. Despite the fact that early studies were hampered by the lack of simple and sensitive methods for the identification of the presence of carbon-to-phosphorus bond in natural samples, it was found to exist in protozoa, bacteria, coelenterates, and mollusks [6–11]. Presumably, the unbreakable record is held by the snail *Helisoma* sp. freshly laid eggs, which contain over 95% of total phosphorus in phosphonate form [12]. Upon embryonic development, phosphonate is converted into phosphoric acid and subsequently incorporated into cellular constituents. It is believed that the physiological role of incorporation of phosphonates into the lipid fraction might function as a means to protect the eggs against predators, because they are presumably not able to disrupt and digest such membranes.

The advent of ^{31}P NMR for the analysis of tissue extracts, body fluids, and later—whole cells provided an effective tool for tracking the forms of phosphorus and its interchanges during organism development and growth. Quite paradoxically, the availability of ^{31}P NMR was accompanied with a significant decrease in the number of papers dealing with distribution of phosphonates in various species. Applications of this simple technique enabled the determination of the presence of C—P bond in bacteria and bacterial communities [13, 14], cyanobacteria [15], sponges [16], higher fungi [17, 18], or even human specimens [19]. However, these studies did not explain if phosphonates are synthesized *de novo* or are introduced to these organisms by cohabiting organisms or diet. On the other hand, phosphonate xenobiotics are quite massively released into environment [20], and various organisms might use them, or products of their decomposition, as building blocks of more complex structures.

Next, gene-based methods for assessing the abundance and identity of biological phosphonate producers were applied. This approach based on knowledge regarding C—P compound biosynthesis. Thus, with a single exception [21], all the known phosphonates are derived from phosphoenolpyruvate by isomerization to phosphonopyruvate in a reaction catalyzed by the phosphoenolpyruvate mutase, followed by its fast utilization because the reaction of formation of the C—P bond is thermodynamically unfavorable (see **Figure 3**). Most common, decarboxylation of phosphonopyruvate by phosphonopyruvate decarboxylase to produce phosphonoacetaldehyde is the next, irreversible step [3, 22–24]. Mining in genome databases for genes related to these two enzymes, as well as their homologs, enabled to determine that 10–15% of bacterial species are able to produce phosphonates [23–25].

Discovery that phosphonates form around 10% of dissolved and particulate phosphorus in the oceans [15, 25, 26] brought the increasing recognition of the importance of these compounds in biogeochemical phosphorus cycling and an awareness of the interdependence between the global phosphorus cycle and those of the other biologically significant elements [27, 28]. It is important because phosphorus availability has been shown to be a key determinant of marine phytoplankton productivity [15]. Phosphonates are mostly concentrated in dissolved organic phosphorus (DOP), an integral and dynamic part of the marine organic matter pool. The composition of the DOP pool is complex and largely unknown, but phosphonates account for one third of its high molecular weight fraction. Thus, they seem to be an important resource of this element for aquatic organisms; however, the understanding of their utilization by eukaryotic phytoplankton is severely limited [29, 30]. They most likely occur in a form of polysaccharides esterified with methylphosphonate (compound **2**, **Figure 1**) and 2-hydroxyethylphosphonate (compound **3**, **Figure 1**). These compounds have been mainly found in *Nitrosopumilus maritimus*, one of the most abundant organisms on the planet and a resident of the oxygen-rich regions of the open oceans [31, 32].

Up to 4% of the methane on Earth comes from the oxygen-rich waters through the cleavage of the highly unreactive carbon-to-phosphorus bond in methyl phosphonate [32]. The production of methylphosphonic acid (MPn) by cyanobacteria or marine archaea related to *N. maritimus* and its subsequent decomposition by phosphate-starved bacterioplankton may partially explain the production of methane in oceanic and lake surfaces [33–35]. The concentration of methane in the upper ocean being above equilibrium with the atmosphere is known as the oceanic methane paradox [36, 38].

Some researchers believe that phosphonates are a form of relic of evolution. Being of slightly lower formal oxidation state, they might predominate in prebiotic reductive conditions [37]. This assumption, although debatable, finds some support by finding several phosphonic acids in Murchison meteorite [39].

3. Ciliatine (AEP, 2-aminoethylphosphonic acid)

Ciliatine (compound **1**) is the most ubiquitous phosphonate present in lower organisms and occurs in remarkably high amounts. It is either presented in a free, unbound form being a common intermediate in numerous phosphonate biosynthetic pathways or incorporated into lipids and glycans. It is not surprising if considering that ciliatine is a formal analog of common component of lipids—phosphoethanolamine (compound **4**). Most of the studies on natural occurrence of ciliatine and its lipids had been published in 1960–1990 and are comprehensively reviewed [2–5]. Only single paper was published after this period. As shown in **Figure 1**, its methylated forms, namely *N*-methyl, *N*,*N*-dimethyl-, and *N*,*N*,*N*-trimethylciliatine (compounds **5**, **6**, and **7**), were also found in lipid fractions of some organisms albeit in significantly smaller quantities. Compound **7** is an analog of the most common component of lipids—phosphocholine (compound **8**). The presence of an unusual aminophosphonate—(R)-2-amino-1-hydroxyethylphosphonic acid (compound **9**) and its acetyl derivative has been determined in lipid fractions of *Bacteriovorax stolpii* [40, 41]. Its configuration was elegantly determined by a combination of chemical synthesis and biochemical studies [42].

Lipids containing aminophosphonates are called phosphonolipids. There are two classes of these compounds—glycerophosphonolipids and sphingophosphonolipids (representative structures are shown in **Figure 2**). They have been isolated from numerous organisms including humans, mammals (sheep, goats, and rats), egg yolk, fish, insects, sea anemones, sponges, numerous species of freshwater and marine mollusks, seeds of plants, protozoa, and bacteria [3, 43–46]. Usually they are a small fraction of the total lipids present, and their isolation and exact identification/characterization are difficult and cumbersome.

The physiologic function of phosphonolipids is still unknown, and the suggested protecting role against predators resulting from their stability toward hydrolysis by lipases and phosphatases has not been proved so far. Moreover, the distribution and abundance of phosphonolipids among organisms vary with species, tissue, or cellular location. For example, vertebrates have sphingophosphonolipids as components of nervous tissue sphingomyelin, while invertebrates frequently contain high levels of these lipids as outer membrane components.

Whereas phosphate is a common modification of polysaccharides, there are only a few examples of polysaccharides containing phosphonate moieties. Their characterization/identification was made possible as well as substantially accelerated by the development of glycomics [47]. Ciliatine and compound **9** have been found to be bound to the sugar moieties of variable glycans (see **Figure 2** for schematic structures). Their occurrence was documented in fractions of glycocerebrosides (lipids) of many lower marine phyla [2, 12, 48], bacterial exopolysaccharides (secreted polysugars into the environment), and outer membrane components

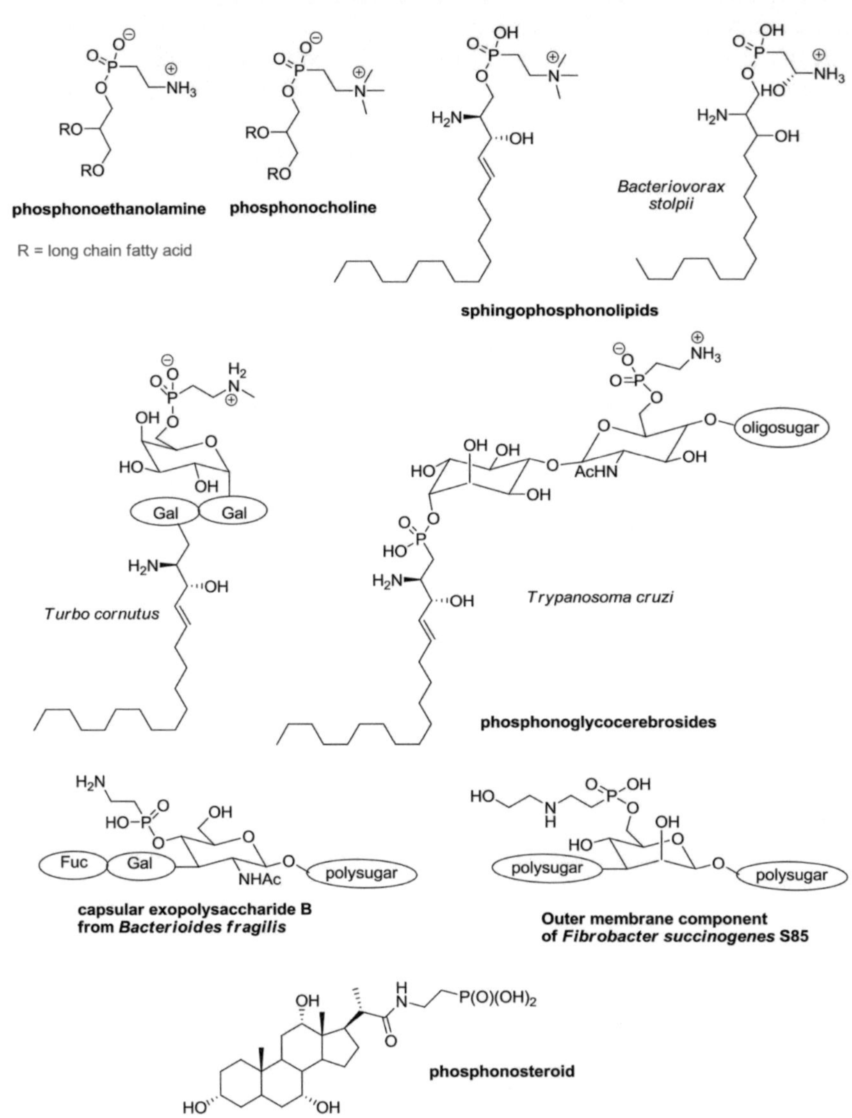

Figure 2.
Representative structures of phosphonolipids, phosphonoglycans, and phosphonosteroid.

[49–51] and glycoproteins deriving from marine snails, common jellyfish and locust [51–54]. Genome scanning led to the identification of methylphosphonic acid (compound **2**) in the exopolysaccharide of the marine archeon *N. maritimus*. Its function is not known, but it is ultimately a major source of methane production by the oceans [32].

Similarly as in the case of phosphonolipids, the physiological role of phosphonoglycans is not known and thus awaits determination. This might be important in the context that the glycans are essential molecules being well known to enable adaptive response to environmental changes [55]. The speculative roles of phosphonoglycans include cell-cell signaling or their action as phosphorus reservoirs in the environments of low phosphate concentration. The second assumption might be supported by the conservation of phosphonolipids at the expense of phosphodiesters in starved conditions by the oyster *Crassostrea virginica* [56]. Other possibility is demonstrated by the fact that *Bacteroides fragilis*, a part of the normal microbiota of the human colon, produces a capsular polysaccharide complex containing ciliatine, which is directly involved in abscess formation in animal models when bacteria are displaced into the bloodstream [57].

It is also important to mention that the phosphonic analog of taurocholic acid was found in the gall bladders of cows [58]; however, this finding may require additional confirmation.

4. Low-molecular phosphonates metabolically related to ciliatine

Biosynthesis of phosphonates starts from rearrangement of phosphoenolpyruvate (compound **10**) into phosphonopyruvate (compound **11**), a reaction catalyzed by phosphonoenolpyruvate mutase. In this equilibrium process, the thermodynamics favors phosphonoenolpyruvate by a factor of at least 500. Thus, phosphonopyruvate has to be rapidly converted into metabolically useful compounds, most favorably in the irreversible reactions. Consequently, it is a key substrate in the synthesis of ciliatine (compound **1**), phosphonoalalnine (compound **12**), 2-hydroxyethylphosphonic acid (compound **3**), phosphonoacetaldehyde (compound **13**), phosphonomethylmalic acid (compound **14**), and 2-keto-4-hydroxy-5-phosphonopentanoic acid (compound **15**). Most of the enzymes involved in the production of these compounds have been isolated and characterized and comprehensively reviewed [2, 59]. The metabolic relationships between these compounds and their precursor role in the synthesis of phosphonate antibiotics are shown in **Figure 3**.

Low-molecular antibiotics such as fosfomycin (compound **17**) [60], fosfonochlorin (compound **18** produced by several strains of *Fusarium* and *Talaromyces flavus*) [61], nitrilaphos and hydroxynitrilaphos (compounds **19** and **20** found in cultivating media of *Streptomyces*) [62], and herbicidal phosphonothrixin (compound **21**, produced by *Saccharothrix*) [63] might be also considered as low-molecular compounds related to ciliatine.

Only one of them—*fosfomycin* (also known as *Monuril*, *Monurol,* or *Monural*), produced by *Pseudomonas* and *Streptomyces,* has found limited use as therapeutic agent to cure urinary tract infections and diabetic foot [60]. It is an active site directed covalent inactivator of muramyl ligase A, the first enzyme of peptidoglycan synthesis, and causes disruption of bacterial cell wall. Unfortunately, bacteria adapted to be able to open the epoxy ring functionality of fosfomycin, thus resulting in the compound deactivation/degradation of this antibiotic and in the microorganism ability to readily develop drug resistance [64]. Quite interestingly, pathways for the biosynthesis of fosfomycin in *Streptomyces* and *Pseudomonas* are different (see **Figure 3**). This shows that synthesis of natural phosphonates does not have to be normalized; many metabolic pathways are still yet to be discovered.

The separate class is aminophosphonate antibacterial antibiotics possessing an amino group in the gamma position in relation to the phosphonic functional group, namely fosmidomycin (compound **22**), and its derivatives denoted as FR900098 (compound **23**), FR-33289 (compound **24**), and FR32863 (compound **25**), originally isolated from culture broths of *Streptomyces* as well as cyclic SF2312 (compound **26**) isolated from *Micromonospora* sp. [65–67]. Their structures are shown in **Figure 4**.

Fosmidomycin and its homologs are potent inhibitors of 1-deoxy-d-xylulose-5-phosphate reductoisomerase, an essential enzyme of the non-mevalonate pathway of isoprenoid biosynthesis being active against a broad range of enterobacteria,

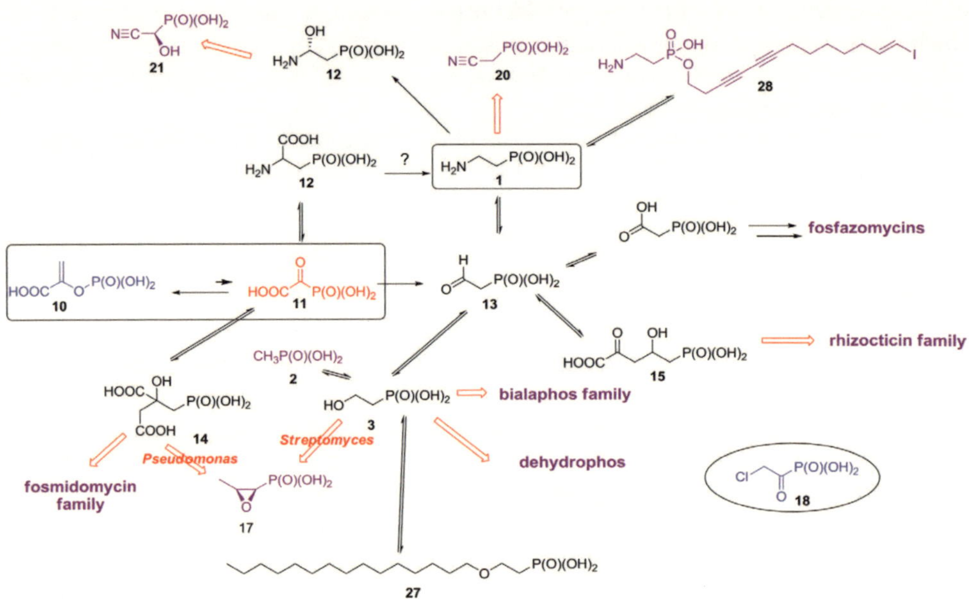

Figure 3.
Metabolic relationship between naturally occurring phosphonates.

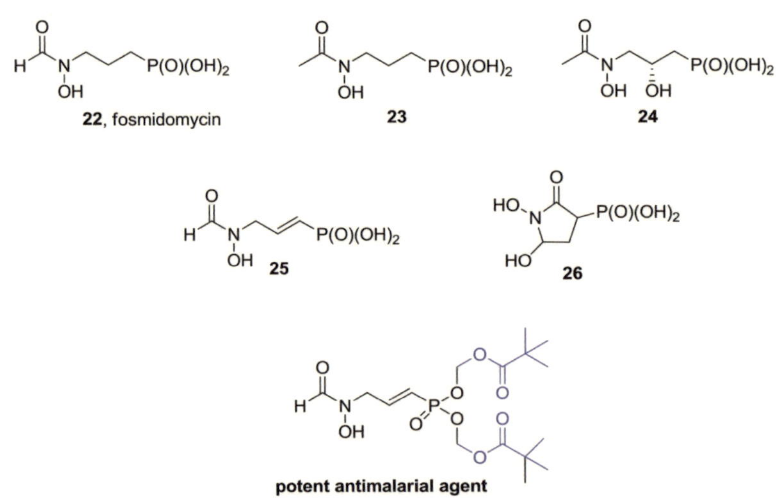

Figure 4.
Antibiotics structurally related to fosmidomycin.

but not against Gram-positive organisms or anaerobes. More importantly, they are blocking the development of isoprenoids in the parasite apicoplast, and thus, structurally modified fosmidomycin derivatives are considered as promising antimalarial agents (for representative structure, see **Figure 4**) [68].

Aphanizomenon flos-aquae is a cyanobacterium that grows in eutrophic Balgavies Loch in Scotland. From its water blooms, a novel biosurfactant of lipidic character—2-acyloxyethylphosphonate (compound **27**) was isolated; however, its ecological function remains to be evaluated [69].

Two unusual placotylene A esters [69] of ciliatine (phosphoiodyn A, compound **28**) and its phosphate congener—phosphoethanolamine (phosphoiodyn B) were isolated from a Korean marine sponge *Placospongia* sp. [70]. Phosphoiodyn A was found to exhibit a potent agonistic activity on human peroxisome proliferator-activated receptor delta (hPPARδ), which is thought to function as an integrator of transcriptional repression and nuclear receptor signaling [16, 71]. This compound, as well as its analogs, demonstrates significant neuroprotective activity in an *in vitro* cellular model indicating that such phosphonates may be an effective novel scaffold for the design of therapeutics for the treatment of neurodegenerative disorders [71].

5. Phosphonopeptide antibiotics

Half of the century after the discovery of ciliatine witnessed a slow progress in the isolation and identification of natural compounds containing the C—P bond with most of them being antibacterials. The majority of these compounds appeared to be peptides containing C-terminal phosphonic acids and mostly differ by their *N*-terminal peptide structure. They have drawn attention not only because of their bioactivity but also because of unusual and interesting chemistry associated with the biosynthesis and biodegradation of these molecules. Structures of antibiotic phosphonopeptides are shown in **Figure 5**.

Bialaphos (compound **29**, [72]) was isolated from as the first such an antibiotic from the culture filtrates of *Streptomyces viridochromogenes* and *Streptomyces hygroscopicus* [72–74]. Further studies indicated that its antibacterial activity is a result of active transport of the peptide across bacterial membrane followed by hydrolysis of the peptide and release of terminal phosphonate—phosphinothricin, which inhibits glutamine synthetase. This enzyme converts glutamic acid and ammonia into glutamine; this reaction is an important step of the nitrogen metabolism in bacteria and plants [75]. That activity of phosphinothricin resulted in its introduction to agriculture as a popular herbicide, and it is sold as ammonium salt under the name glufosinate. Its application causes accumulation of ammonia in plants and consequently plant death [76]. It is worth to notice that bialaphos also exerts herbicidal activity and was applied in Japan [77]. Its activity relays on hydrolysis of bialaphos in plant tissues and release of herbicidal phosphinothricin. Further studies on bialaphos resulted in isolation of tetrapeptide trialaphos (compound **30**) [78] and phosalacine (compound **31**) [79] both of the same mechanism of action. Finally, studies on biosynthesis of this compound resulted in the identification of its desmethyl analog **32**, which is an intermediate in bialaphos metabolism.

The antibacterial activity of bialaphos is typical for all the phosphonopeptides. Peptide parts of these antibiotics usually function as a targeting unit. Thus, the peptides are efficiently transported through bacterial (or fungal) membranes and after hydrolysis release phosphonic acid, which exerts its toxic action by inhibiting parasite vital enzymes—in this case glutamine synthetase. This mechanism of action is shown schematically in **Figure 6**.

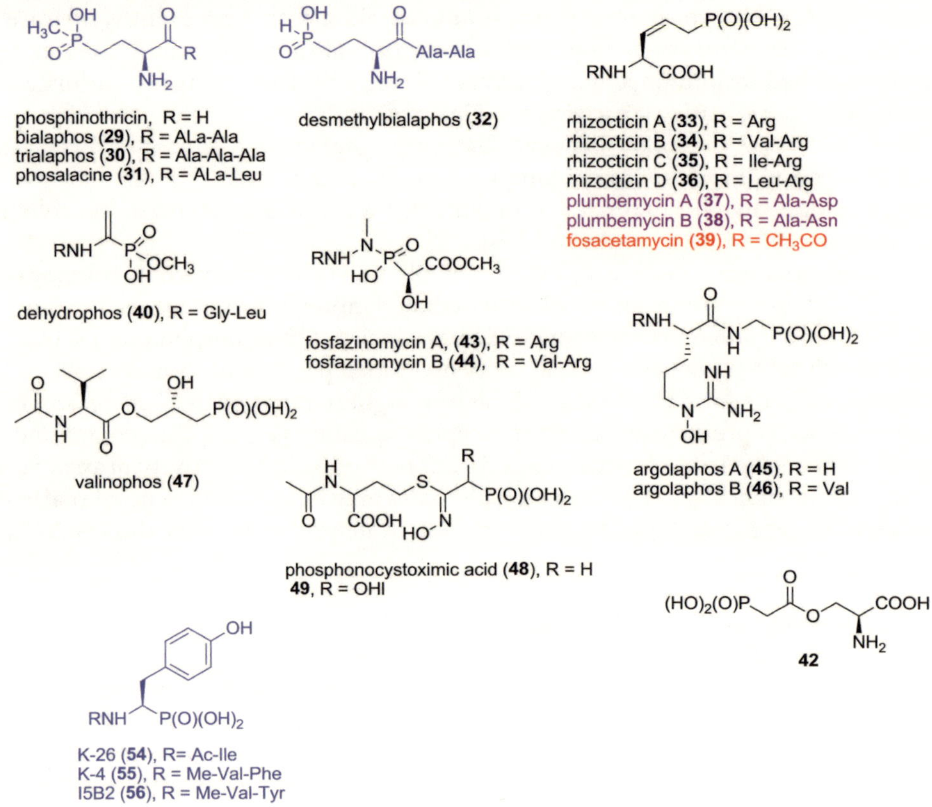

Figure 5.
Phosphonopeptide antibiotics.

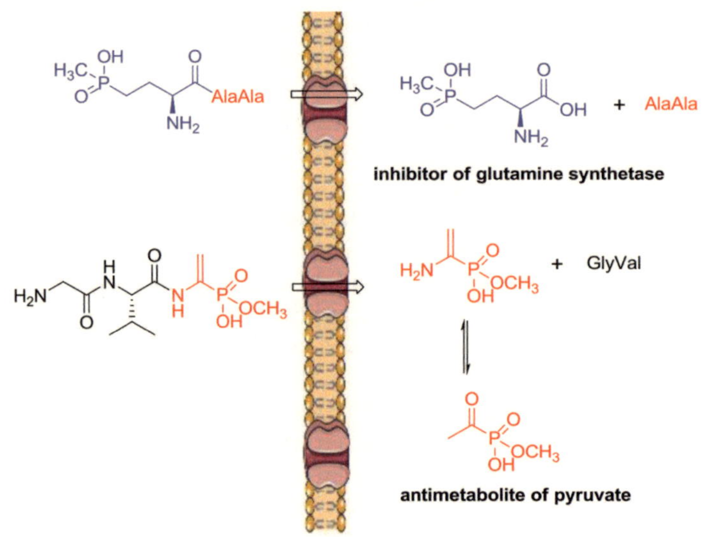

Figure 6.
Representative mechanism of action of phosphonopeptides.

The following years brought the discovery of a family of antibiotics called rhizocticins (compounds **33–36**) [80, 81], plumbemycins (compounds **37** and **38**) [81–83], and phosacetamycin (compound **39**) [84], first isolated as secondary metabolites of *Bacillus subtilis* on the basis of their antifungal activity and were later found as products of *Streptomyces plumbeus*. They form a library of di- and tripeptides containing C-terminal (*Z*)-*L*-2-amino-5-phosphono-3-pentenoic acid, a mimetic of phosphonothreonine, which is the substrate for threonine synthetase. Thus, after the release from the peptide aminophosphonate acts as a potent inhibitor of this enzyme [85].

Dehydrophos (compound **40**) was first isolated from the broth of *Streptomyces luridus* as a broad-spectrum antibiotic affective in chicken model of *Salmonella* infection [86]. The history of determination of its structure is rather long and led to three propositions of which the last one appeared to be reasonable and compelling. It is a dehydrophosphonopeptide, which, after the cleavage of the peptide bond, provides an analog of dehydroalanine, which is then converted into methyl acetylphosphonate (compound **41**, an analog of pyruvic acid), which is strongly antibacterial by acting most likely as antimetabolite of pyruvate (**Figure 6**) [87]. Thus, it was considered as a lead compound for the design of novel antibacterial agents [88]. The non-typical and innovative is the application of its biosynthetic enzymes for obtaining new antibacterial phosphonopeptides [89]. Recently, the role of nonribosomal peptidyl transferase DhpH in the formation of peptide bond in dehydrophos was studied in detail using phosphonic analog of alanine and various amino acid-tRNAs as substrates [90].

Phosphonopeptides have very limited utility in human medicine because they are readily hydrolyzed in body fluids and released aminophosphonic acids that are not able to cross bacterial or fungal cell barriers and to exert antibiotic action. Additionally, they are being readily excreted through urine.

Published in 2015 work of Metcalf and van der Donk brought a significant breakthrough in studies on naturally occurring phosphonate antibiotics. By a clever combination of the mining of the genome of 10,000 of actinomycetes and selective labeling of phosphonate metabolites, they rediscovered a large number of old phosphonates and discovered 19 new compounds [24]. This opened a genetic approach in natural phosphonate chemistry and biochemistry, especially enabling the identification of metabolic pathways leading to this class of compounds. An important and instructive example here is an activation of gene cluster from *Streptomyces* sp. NRRL F-525 and its reengineering in *Streptomyces lividans*, which resulted in the isolation of *O*-phosphonoacetic acid serine (compound **42**) [91].

One of the examples of rediscovered compounds is fosfazinomycins A and B (compounds **43** and **44**), identified 30 years after their original isolation from *Streptomyces lavendofoliae* and *Streptomyces unzenensis* [92, 93]. They are a very specific since they contain an exotic structural feature, which is the hydrazide linkage between the carboxylic acid of peptidyl arginine and the phosphonic acid. Fosfazinomycin was also found further in one of 210 substances present in 42 actinomycetes associated with the Baltic sponge *Halichondria panacea* [94].

The genetic approach also enabled the isolation and characterization of novel of *Streptomyces* peptidomimetics such as argolaphos A and B (compounds **45** and **46**) and valinophos (compound **47**) [24]. Similar approach was used for the isolation of phosphonocystoximate and its hydroxylated derivative (compounds **48** and **49**) [24]. Detailed NMR studies on their biosynthesis, which starts from ciliatine and its analog—compound **9**, enabled to confirm the presence of intermediates such as mixtures of the (E)- and (Z)-isomers of corresponding oximes (compounds **50** and **51**), substrates for the synthesis of phosphonocystoximate and its hydroxylated derivative [95]. They are formed by the action of specific flavin-dependent, oxime-forming N-oxidases. These oxidases are also able to convert the oximes **50** and **51** into corresponding nitroethylphosphonates (compounds **52** and **53**) [96]. Structures of these intermediates and side products are depicted in **Figure 7**.

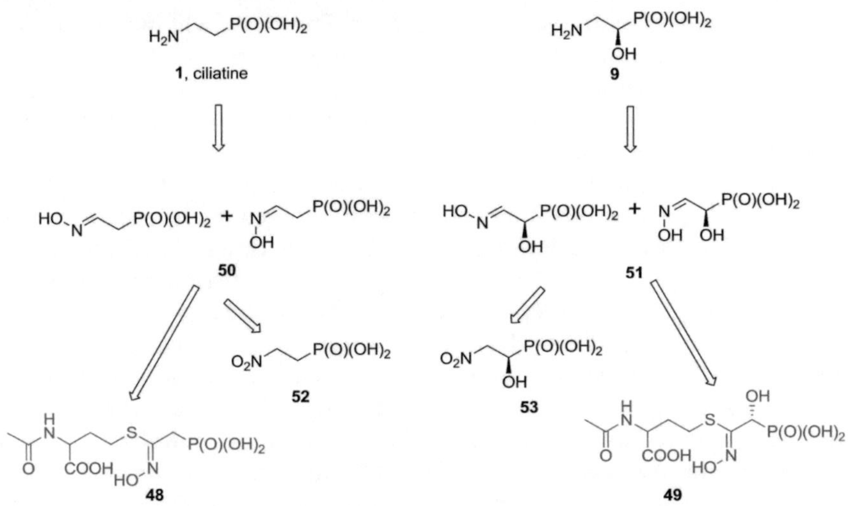

Figure 7.
Intermediates and side products in the synthesis of phosphonocystoximate.

A separate group of phosphonic peptidomimetics is compounds denoted as K-26, K4, and I5B2 (compounds **54, 55,** and **56,** respectively) [21, 97–99], a small family of bacterial secondary metabolites, tripeptides terminated by an unusual phosphonate analog of tyrosine (see **Figure 6**). They are produced by three different actinomycetales and act as potent inhibitors of human angiotensin-I converting enzyme selectively targeting the eukaryotic family of the enzyme [100, 101]. These compounds derived from *L*-tyrosine, which suggests the existence of novel and not discovered yet reaction of carbon-to-phosphorus bond formation [21, 102].

6. Conclusions

Natural phosphonates might be considered as simple analogs of phosphate esters and/or carboxylic acids. The inherent stability of the C—P bond causes that they often display promising activities as enzyme inhibitors and therefore might be considered as drugs or agrochemicals. Moreover, the wide use of xenobiotics containing carbon-to-phosphorus bond has led to the spread of these compounds in the environment, which may result in their incorporation into variable metabolic pathways. All of this stimulate interest in these, still somewhat exotic, compounds. The development of ^{31}P NMR and genomics supplemented by biochemical studies resulted in the development of new detection technologies, which enormously speed out the discovery of novel naturally occurring phosphonates, identification of their metabolic pathways (both biosynthesis and degradation), and their use as lead compounds for the design of new promising medicines. With the exception of the identification of antibacterial and antifungal antibiotics, these studies are not accompanied, however, with the determination of physiologic importance of these compounds.

Acknowledgements

This work was supported by statuary grants of Wrocław University of Science and Technology and National Science Centre, Poland (grant 2016/21/B/ST5/00115).

Conflict of interest

I declare that there is no conflict of interest that might have any bearing on research reported in this work.

Author details

Paweł Kafarski
Department of Bioorganic Chemistry, Faculty of Chemistry, Wrocław University of Science and Technology, Wrocław, Poland

*Address all correspondence to: pawel.kafarski@pwr.edu.pl

References

[1] Horiguchi M, Kandatsu M. Isolation of 2-aminoethane phosphonic acid from rumen protozoa. Nature. 1959;**184**:901-902. DOI: 10.1038/184901b0

[2] Horsman GP, Zechel DL. Phosphonate biochemistry. Chemical Reviews. 2017;**117**:5704-5783. DOI: 10.1021/acs.chemrev.6b00536

[3] Ju K-S, Doroghazi JR, Metcalf WW. Genomics-enabled discovery of phosphonate natural products and their biosynthetic pathways. Journal of Industrial Microbiology & Biotechnology. 2014;**421**:345-356. DOI: 10.1007/s10295-013-1375-2

[4] Mastalerz P, Kafarski P. Naturally occurring aminophsophonic and aminophosphinic acids. In: Kukhar VP, Hudson HR, editors. Aminophosphonic and Aminophosphinic Acids. Chichester: Wiley; 2000. pp. 1-31

[5] Petkowski JJ, Bains W, Seager S. Natural products containing "rare" organophosphorus functional groups. Molecules. 2019;**24**:866. DOI: 10.3390/molecules24050866

[6] Kitterdge JS, Roberts E. A carbon-phosphorus bond in nature. Science. 1969;**164**:37-42. DOI: 10.1126/science.164.3875.37

[7] Horiguchi M. Occurrence, identification and properties of phosphonic and phosphinic acids. In: Hori T, Horiguchi M, Hayashi M, editors. Biochemistry of Natural C-P Compounds. Shiga: Japanese Association for Research on the Biochemistry of C-P Compounds; 1984. pp. 24-52

[8] Tamari M, Kametaka M. Isolation and identification of ciliatine (2-aminoethylphosphonic acid) from phospholipids of the oyster, *Crassostrea gigas*. Agricultural and Biological Chemistry. 1972;**36**:1147-1152. DOI: 10.1271/bbb1961.36.1147

[9] Kittredge JS, Roberts E, Simonsen DG. The occurrence of free 2-aminoethylphosphonic acid in the aea anemone, *Anthopleura elegantissima*. Biochemistry. 1962;**1**:624-628. DOI: 10.1021/bi00910a013

[10] Liang CR, Rosenberg H. On the distribution and biosynthesis of 2-aminoethylphosphonate in two terrestrial molluscs. Comparative Biochemistry and Physiology. 1968;**25**:673-681. DOI: 10.1016/0010-406X(68)90377-0

[11] Kariotoglou DM, Mastronicolis SK. Sphingophosphonolipid molecular species from edible mollusks and a jellyfish. Comparative Biochemistry and Physiology. B. 2003;**136**:27-44. DOI: 10.1016/S1096-4959(03)00168-4

[12] Miceli MV, Henderson TO, Myers TC. Alkylphosphonic acid distribution in the planorbid snail *Helisoma* sp. Comparative Biochemistry and Physiology. B. 1987;**2**:603-611. DOI: 10.1016/0305-0491(87)90351-8

[13] Jayasimhulu K, Hunt SM, Kaneshiro ES. Detection and identification of *Bacteriovorax stolpii* UKi2 sphingophosphonol;ipid molecular species. Journal of the American Society for Mass Spectrometry. 2007;**18**:394-403. DOI: 10.1016/j.jasms.2006.10.014

[14] Turner BL, Baxter R, Mathieu N, Sjögersten S, Whitton B. Phosphorus compounds in subartic Fennoscandian soils at the mountain birch (*Betula pubescens*)-Tundra ecotone. Soil Biology and Biochemistry. 2004;**36**:815-823. DOI: 10.1016/j.soilbio.2004.01.011

[15] Dyhrman ST, Benitez-Nelson CR, Orchard ED, Haley ST, Pellechia PJ. A microbial source of phosphonates

in oligotrophic marine systems. Nature Geoscience. 2009;**2**:699. DOI: 10.1038/ngeo639

[16] Kim H, Chin J, Choi H, Baek K, Lee TG, Park SE, et al. Phosphoiodyns A and B, unique phosphorus-containing iodinated polyacetylenes from a Korean sponge *Placospongia* sp. Organic Letters. 2013;**15**:100-103. DOI: 10.1021/ol3031318

[17] Koukol O, Novák F, Hrabal R. Composition of the organic phosphorus fraction in basidiocarps of saprotrophic and mycorrhizal fungi. Soil Biology and Biochemistry. 2008;**40**:2464-2467. DOI: 10.1016/j.soilbio.2008.04.021

[18] Maciejczyk E, Wieczorek D, Zwyrzykowska A, Halama M, Jasicka-Misiak I, Kafarski P. Phosphorus profile of basidomycetes. Phosphorus, Sulfur and Silicon and the Related Elements. 2015;**190**:763-768. DOI: 10.1080/10426507.2014.99

[19] Glonek T, Henderson TO, Hilderbrand RL, Myers TC. Biological phosphonates: Determination by phosphorus-31 nuclear magnetic resonance. Science. 1970;**169**:172-174. DOI: 10.1126/science.169.3941.192

[20] Studnik H, Liebsch S, Forlani G, Wieczorek D, Kafarski P, Lipok J. Aminopolyphosphonates—Chemical features and practical uses, environmental durability and biodegradation. New Biotechnology. 2015;**32**:1-6. DOI: 10.1016/j.nbt.2014.06.007

[21] Ntai I, Manier ML, Hachey DL, Bachmann BO. Biosynthetic origins of C-P bond containing tripeptide K-26. Organic Letters. 2005;7:2763-2765. DOI: 10.1021/ol051091d

[22] Peck SC, van der Donk WA. Phosphonate biosynthesis and catabolism: A treasure trove of unusual enzymology. Current Opinion in Chemical Biology. 2013;**19**:580-588. DOI: 10.1016/j.cbpa.2013.06.018

[23] Villarreal-Chiu JF, Quinn JP, McGrath JW. The genes and enzymes of phosphonate metabolism by bacteria, and their distribution in the marine environment. Frontiers in Microbiology. 2012;**3**:art.19. DOI: 10.3389/fmicb.2012.00019

[24] Yu JX, Doroghazi JR, Janga SC, Zhang JK, Circello B, Grifiinj BM, et al. Diversity and abundance of phosphonate biosynthetic genes in nature. Proceedings of the National Academy of Sciences of the United States of America. 2013;**110**:20759-20764. DOI: 10.1073/pnas.1315107110

[25] Scott KM, Sievert SM, Abril FN, Ball LA, Barrett CJ, Blake RA, et al. The genome of deep-see vent chemolithoautitroph *Thiomicrospira crunogena* XCL-2. PLoS Biology. 2006;**4**:e383. DOI: 10.1371/journal.pbio.0040383

[26] Whitney LP, Lomas LW. Phosphonate utilization by eukaryotic phytoplankton. Limnology and Oceanography Letters. 2019;**4**:18-24. DOI: 10.1002/lol2.10100

[27] Chin JP, McGrath JW, Quinn JP. Microbial transformations in phosphonate biosynthesis and catabolism, and their importance in nutrient cycling. Current Opinion in Chemical Biology. 2016;**31**:50-57. DOI: 10.1016/j.cbpa.2016.01.010

[28] Lüke C, Speeth DR, Kox MAR, Villanueva M, Jetten MSM. Metagenomic analysis of nitrogen and methane cycling in the Arabian Sea oxygen minimum zone. PeerJ. 2016;**4**:e1924. DOI: 10.7717/peerj.1924

[29] Born DA, Ulrich EC, Ju K-S, Peck S, van der Donk WA, Drennan CL. Structural basis for

methylphosphonate biosynthesis. Science. 2017;**358**:1336-1339. DOI: 10.1126/science.aao3435

[30] Nowack B. Environmental chemistry of phosphonates. Water Resources. 2003;**37**:2533-2546. DOI: 10.1016/S0043-1354(03)00079-4

[31] Benitez-Nelson CR, O'Neill L, Kolowith LC, Pellecia P, Thunell R. Phosphonates and particulate organic phosphorus cycling in an anoxic marine basin. Limnology and Oceanography. 2004;**49**:1593-1604. DOI: 10.4319/lo.2004.49.5.1593

[32] Metcalf WW, Griffin BM, Cicchillo RM, Gao J, Chandra Janga S, Cooke HA, et al. Synthesis of methylphosphonic acid by marine microbes: A source for methane in the aerobic ocean. Science. 2012;**337**:1104-1107. DOI: 10.1126/science.1219875

[33] del Valle DA, Karl DM. Aerobic production of methane from dissolved water-column methylphosphonate and sinking particles in the North Pacific subtropical gyre. Aquatic Microbial Ecology. 2014;**73**:93-105. DOI: 10.3354/ame01714

[34] Bižić-Ionescu M, Klintzsch T, Ionescu D, Hindiyeh MY, Günthel M, Muro-Pastor AM, et al. Widespread methane formation by Cyanobacteria in aquatic and terrestrial ecosystems. BioRxiv. DOI: 10.1101/398958

[35] Beversdorf LJ, White AE, Björkman KM, Letelier RM, Karl DM. Phosphonate metabolism of *Trichodesmium* IMS101 and the production of greenhouse gases. Limnology and Oceanography. 2010;**55**:1768-1778. DOI: 10.4319/lo.2010.55.4.1768

[36] Repeta DJ, Ferrón S, Sosa OA, Johnson CG, Repeta LD, Acke M, et al. Marine methane paradox explained by bacterial degradation of dissolved organic matter. Nature Geoscience. 2016;**9**:884-887. DOI: 10.1038/ngeo2837

[37] McGrath JW, Chin JP, Quinn JP. Organophosphonate reveals: New insights into microbial metabolism of ancient molecules. Nature Reviews. Microbiology. 2013;**11**:411-419. DOI: 10.1038/nrmicro3011

[38] Schwartz AW. Phosphorus in prebiotic chemistry. Philosophical Transactions of the Royal Society B. 2006;**361**:1743-1749. DOI: 10.1098/rstb.2006.1901

[39] Cooper GW, Onwo WM, Cronin JR. Alkyl phosphonic acids and sulfonic acids in the Murchison meteorite. Geochimica et Cosmochimica Acta. 1992;**56**:4109-4115. DOI: 10.1016/0016-7037(92)90023-C

[40] Jayasimhulu K, Hunt SM, Kaneshiro ES, Watanabe Y, Giner J-L. Detection and identification of *Bacteriovorax stolpii* UKi2 sphingophosphonolipid molecular species. Journal of the American Society for Mass Spectrometry. 2007;**18**:394-403. DOI: 10.1016/j.jasms.2006.10.014

[41] Watanabe Y, Nakayima N, Hoshino T, Jayasimhulu K, Brooks EE, Kaneshiro ES. A novel sphingophosphonolipid head group 1-hydroxy-2-aminoethyl phosphonate in *Bdellovibrio stolpii*. Lipids. 2001;**36**:513-519. DOI: 10.1007/s11745-001-0751-3

[42] Pallitsch K, Happl B, Stieger C. Determination of the absolute configuration of (−)-hydroxynitrilaphos and related biosynthetic questions. Chemistry-A European Journal. 2017;**23**:15655-15665. DOI: 10.1002/chem.201702904

[43] Moschidis MC. Phosphonolipids. Progress in Lipid Research. 1984;**23**:223-246. DOI: 10.1016/0163-7827(84)90012-2

[44] Muralidhar P, Radhika P, Krishna N, Venkata Rao D, Bheemasankara Rao CH. Sphingolipids from marine organisms: A review. Natural Product Sciences. 2009;**9**:117-142

[45] Satake M, Miyamoto E. A group of glycosphingolipids found in an invertebrate: Their structures and biological significance. Proceedings of the Japan Academy, Series B. 2012;**88**:509-517. DOI: 10.2183/pjab.88.509

[46] Mukhamedova KS, Glushenkova AI. Natural phosphonolipids. Chemistry of Natural Compound. 2000;**36**:329-341. DOI: 10.1023/A:1002804409503

[47] Paschinger K, Wilson IB. Analysis of zwitterionic and anionic *N*-linked glycans from invertebrates and protists by mass spectrometry. Glycoconjugate Journal. 2016;**33**:273-283. DOI: 10.1007/s10719-016-9650-x

[48] Korn ED, Dearborn DG, Fales HM, Sokolowski EA. Phosphonoglycan. A major polysaccharide constituent of the amoeba plasma membrane contains 2-aminoethylphosphonic acid and 1-hydroxy-2-aminoethylphosphonic acid. The Journal of Biological Chemistry. 1973;**248**:2257-2259

[49] Vinogradov E, Egbosimba EE, Perry MB, Lam JS, Forsberg CW. Structural analysis of the carbohydrate components of the outer membrane of the lipopolysaccharide-lacking cellulolytic ruminal bacterium *Fibrobacter succinogenes*. European Journal of Biochemistry. 2001;**268**:3566-3576. DOI: 10.1046/j.1432-1327.2001.02264.x

[50] Baumann H, Tzianabos AO, Brisson JR, Kasper DL, Jennings HJ. Structural elucidation of two capsular polysaccharides from one strain of *Bacteroides fragilis* using high-resolution NMR spectroscopy. Biochemistry. 1992;**31**:4081-4089. DOI: 10.1021/bi00131a026

[51] Young NM, Foote SJ, Wakarchuk WW. Review of phosphocholine substituents on bacterial pathogen plycans: Synthesis, structures and interactions with host proteins. Molecular Immunology. 2013;**56**:563-573. DOI: 10.1016/j.molimm.2013.05.237

[52] Urai M, Nakamura T, Uzawa J, Baba T, Taniguchi K, Seki H, et al. Structural analysis of *O*-glycans of mucin from jellyfish (*Aurelia aurita*) containing 2-aminoethylphosphonate. Carbohydrate Research. 2009;**344**:2182-2187. DOI: 10.1016/j.carres.2009.08.001

[53] Hård K, Van Doorn JM, Thomas-Oates JE, Kamerling JP, Van der Horst DJ. Structure of the Asn-linked oligosaccharides of Apolipophorin III from the insect *Locusta migratoria*. Carbohydrate-linked 2-aminoethylphosphonate as a constituent of a glycoprotein. Biochemistry. 1993;**32**:766-775. DOI: 10.1021/bi00054a005

[54] Eckmair B, Jin C, Abed-Avandi D, Paschinger K. Multistep fractionation and mass spectrometry reveal zwitterionic and anionic modifications of the N- and O-glycans of a marine snail. Molecular & Cellular Proteomics. 2016;**15**:573-597. DOI: 10.1074/mcp.M115.051573

[55] Lauc G, Krištić J, Zoldoš V. Glycans—The third revolution in evolution. Frontiers in Genetics. 2014;**5**:art.145. DOI: 10.3389/fgene.2014.00145

[56] Swift ML. Phosphono-lipid content of the oyster, *Crassostrea virginica*, in three physiological conditions. Lipids. 1977;**12**:449-451. DOI: 10.1007/BF02533632

[57] Onderdonk AB, Kasper DL, Cisneros RL, Bartlett JG. The capsular

polysaccharide of *Bacteroides fragilis* as a virulence factor: Comparison of the pathogenic potential of encapsulated and unencapsulated strains. The Journal of Infectious Diseases. 1977;**136**:82-89. DOI: 10.1093/infdis/136.1.82

[58] Tamari M, Ogawa M, Kametaka M. A new bile acid conjugate, ciliatocholic acid, from bovine gall bladder bile. Journal of Biochemistry. 1976;**80**:371-377. DOI: 10.1093/oxfordjournals.jbchem.a131286

[59] Peck SC, van der Donk W. Phosphonate biosynthesis and catabolism: A treasure trove for unusual enzymology. Current Opinion in Chemical Biology. 2013;**17**:580-588. DOI: 10.1016/j.chpa.2013.06.018

[60] Falagas ME, Vouloumanou K, Samonis G, Vardakas KZ. Fosfomycin. Clinical Microbiology Reviews. 2016;**29**:321-347. DOI: 10.1128/CMR.00068-15

[61] Takeuchi M, Nakajima M, Ogita T, Inukai K, Kodama K, Furuya K, et al. Fosfonochlorin, a new antibiotic with spheroplast forming activity. Journal of Antibiotics. 1989;**42**:198-205. DOI: 10.7164/antibiotics.42.198

[62] Cioni JP, Doroghazi JR, Ju JR, Yu K-S, Evans BS, Lee J, et al. Cyanohydrin phosphonate natural products from *Streptomycin regenesis*. Journal of Natural Products. 2014;7:243-249. DOI: 10.1021/np400722m

[63] Takahashi E, Kimura T, Nakamura K, Arahira M, IIda M. Phosphonothrixin, a novel herbicidal antibiotic produced by Saccharothrix sp. ST-888. I. Taxonomy, fermentation, isolation and biological properties. Journal of Antibiotics (Tokyo). 1995;**48**:1124-1129. DOI: 10.7164/antibiotics.48.1124

[64] Silver LL. Fosfomycin: Mechanism and resistance. Cold Spring Harbor Perspectives in Medicine. 2017;7:a025262. DOI: 10.1101/cshperspect.a025262

[65] Iguchi E, Okuhara M, Koshaka M, Aoki H, Imanaka H. Studies on new phosphonic acid antibiotics. II. Taxonomic studies on producing organisms of the phosphonic acid and related compounds. Journal of Antibiotics. 1980;**33**:19-23. DOI: 10.7164/antibiotics.33.18

[66] Ohba K, Sato Y, Sasaki T, Sezaki M. Studies on a new phosphonic acid antibiotic SF2312.II isolation, physico-chemical properties and structure. Science Reports. 1986;**25**:18-22

[67] Okuhara M, Kuroda Y, Goto T, Okamoto M, Terano H, Kohsaka M, et al. Studies on a new phosphonic acid antibiotic III. Isolation and characterisation of FRFR-31564, FR-32863 and FR-33289. Journal of Antibiotics. 1980;**33**:24-28. DOI: 10.7164/antibiotics.33.24

[68] Edwards RL, Brothers RC, Wang X, Maron MI, Ziniel PD, Tsang PS, et al. MEPicides: Potent antimalarial prodrugs targeting isoprenoid biosynthesis. Scientific Reports. 2017;7:art.8400

[69] Kaya K, Morrison LF, Codd GA, Metcalf JS, Sano T, Takagi H, et al. A novel biosurfactant, 2-Acyloxyethylphosphonate, isolated from Waterblooms of *Aphanizomenon flos-aquae*. Molecules. 2006;**11**:539-548. DOI: 10.3390/11070539

[70] Kim H, Kim K-J, Jeon J-T, Kim S-H, Won DH, Choi H, et al. Placotylene A, an inhibitor of the receptor activator of nuclear factor-κB ligand-induced osteoclast differentiation, from a Korean sponge *Placospongia* sp. Molecules. 2014;**12**:2054-2065. DOI: 10.3390/d12042054

[71] Kinarivala N, Suh JH, Botros M, Webb P, Trippier PC. Pharmacophore

elucidation of phosphoiodyn A—Potent and selective peroxisome proliferator-activated receptor β/δ agonists with neuroprotective activity. Bioorganic & Medicinal Chemistry Letters. 2016;**26**:1889-1893. DOI: 10.1016/j.bmcl.2016.03.028

[72] Ogawa Y, Tsuruoka T, Inouye S, Niida T. Studies on a new antibiotic SF-1293. II. Chemical structure of antibiotic SF-1293. Science Reports. 1973;**13**:42-48

[73] Bayer E, Gugel KH, Hägele K, Hagenmaier H, Jessipow S, König WA, et al. Metabolic products of microorganisms. 98. Phosphinothricin and phosphinothricyl-alanyl-alanine. Helvetica Chimica Acta. 1972;**55**:224-239. DOI: 10.1002/hlca.19720550126

[74] Blodgett JAV, Zhang JK, Yu X, Metcalf WW. Conserved biosynthetic pathways for phosalacine, bialaphos and newly discovered phosphonic acid natural products. The Journal of Antibiotics. 2016;**69**:15-25. DOI: 10.1038/ja.2015.77

[75] Wild A, Ziegler C. The effect of bialaphos on ammonium-assimilation and photosynthesis I. Effect on the enzymes of ammonium-assimilation. Zeitschrift für Naturforschung. 1984;**44**:97-102. DOI: 10.1515/znc-1989-1-217

[76] Devkota P, Johnson WG. Glufosinate efficacy as influenced by carrier water pH, hardness, foliar fertilizer, and ammonium sulfate. Weed Technology. 2016;**30**:848-859. DOI: 10.1614/WT-D-16-00053.1

[77] Tachibana T, Kaneko T. Development of an herbicide, bialaphos. Journal of Pest Science. 1986;**11**:297-304. DOI: 10.1584/jpestics.11.297

[78] Kato H, Nagayama K, Abe H, Kobayashi R, Ishihara E. Isolation, structure and biological activity of trialaphos. Agricultural and Biological Chemistry. 1991;**55**:1133-1134. DOI: 10.1080/00021369.1991.10870694

[79] Omura S, Murata M, Hanaki H, Hinotozawa K, Oiwa R, Tanaka H. Phosalacine, a new herbicidal antibiotic containing phosphinothricin. Fermentation, isolation, biological activity and mechanism of action. The Journal of Antibiotics. 1984;**37**:829-835. DOI: 10.7164/antibiotics.37.829

[80] Rapp C, Jung G, Kugler M, Loeffler W. Rhizocticins—New phosphono-oligopeptides with antifungal activity. European Journal of Organic Chemistry. 1988:655-661. DOI: 10.1002/jlac.198819880707

[81] Gahungu M, Arguelles-Arias A, Fickers P, Zervosen A, Joris B, Damblon C, et al. Synthesis and biological evaluation of potential threonine synthase inhibitors: Rhizocticin A and Plumbemycin A. Bioorganic & Medicinal Chemistry. 2013;**21**:4958-4967. DOI: 10.1016/j.bmc.2013.06.064

[82] Park BK, Hirota A, Sakai H. Studies on new antimetabolite produced by microorganism. III. Structure of plumbemycin A and B, antagonists of L-threonine from *Streptomyces plumbeus*. Agricultural and Biological Chemistry. 1977;**41**:573579. DOI: 10.1271/bbb1961.41.573

[83] Borisova SA, Circello BT, Zhang JK, van der Donk WW, Metcalf WW. Biosynthesis of rhizocticins, antifungal phosphonate oligopeptides produced by *Bacillus subtilis* ATCC6633. Chemistry & Biology. 2010;**17**:28-37. DOI: 10.1016/j.chembiol.2009.11.017

[84] Evans BS, Zhao C, Gao J, Evans CM, Ju K-S, Doroghazi JR, et al. Discovery of the antibiotic phosacetamycin via a new mass spectrometry-based method for phosphonic acid detection. ACS

Chemical Biology. 2013;**8**:908-913. DOI: 10.1021/cb400102t

[85] Kugler M, Loeffler W, Rapp C, Kern A, Jung G. Rhizocticin A, an antifungal phosphono-oligopeptide of *Bacillus subtilis* ATCC 6633: Biological properties. Archives of Microbiology. 1990;**153**:276-281. DOI: 10.1007/BF00249082

[86] Johnson RD, Kastner RM, Larsen SH, Ose EE. Antibiotic A53868 and process for production thereof. U.S. patent 4,482,488; 1988

[87] Circello BT, Miller CG, Lee J-H, van der Donk WA, Metcalf WW. The antibiotic dehydrophos is converted to a toxic pyruvate analog by peptide bond cleavage in *Salmonella enterica*. Antimicrobial Agents and Chemotherapy. 2011;**55**:3357-3362. DOI: 10.1128/AAC.01483-10

[88] Jiménez-Andreu MM, Quintana AL, Aínsa JA, Sayago FJ, Cativiela C. Synthesis and biological activity of dehydrophos derivatives. Organic & Biomolecular Chemistry. 2019;**5**:1097-1112. DOI: 10.1039/c8ob03079k

[89] Bougioukou DJ, Ting CP, Peck SC, Mukherjee S, van der Donk WA. Use of the dehydrophos biosynthetic enzymes to prepare antimicrobial analogs of alaphosphin. Organic & Biomolecular Chemistry. 2019;**4**:822-829. DOI: 10.1039/C8OB02860E

[90] Bougioukou DJ, Mukherjee S, van der Donk WA. Revisiting the biosynthesis of dehydrophos reveals a tRNA-dependent pathway. Proceedings of the National Academy of Sciences of the United States of America. 2013;**110**:10952-11057. DOI: 10.1073/pnas.1303568110

[91] Freestone TS, Ju K-S, Wang B, Zhao H. Discovery of a phosphonoacetic acid derived natural product by pathway refactoring. ACS Synthetic Biology. 2017;**6**:217-223. DOI: 10.1021/acssynbio.6b00299

[92] Gunji S, Arima K, Beppu T. Screening of antifungal antibiotics according to activities inducing morphological abnormalities. Agricultural and Biological Chemistry. 1983;**41**:2016-2069. DOI: 10.1080/00021369.1983.10865911

[93] Gao J, Ju K-S, Yu X, Velásques JE, Mukherjee S, Lee J, et al. Use of phosphonate methyltransferase in the identification of the fosfazinomycin biosynthetic gene cluster. Angewandte Chemie, International Edition. 2014;**53**:1334-1337. DOI: 10.1002/anie.201308363

[94] Schneemann I, Nagel K, Labes A, Wiese J, Imhoff JF. Comprehensivei of marine *Actinobacteria* associated with the sponge *Halichondria panacea*. Applied and Environmental Microbiology. 2010;**76**:3702-3714. DOI: 10.1128/AEM.00780-10

[95] Goettge MN, Cioni JP, Ju K-S, Pallitsch K, Metcalf WW. PcxL and HpxL are flavin-dependent, oxime-forming N-oxidases in phosphonocystoximic acid biosynthesis in *Streptomyces*. The Journal of Biological Chemistry. 2018;**293**:6859-6868. DOI: 10.1074/jbc.RA118.001721

[96] Pallitsch K, Kalina T, Stanković T. Synthetic phosphonic acids as potent tools to study phosphonate enzymology. Synlett. 2019;**30**:770-776. DOI: 10.1055/s-0037-1611460

[97] Yamato M, Koguchi T, Okachi R, Yamada K, Nakayama K, Kase H, et al. K-26, a novel inhibitor of angiotensin I converting enzyme produced cy an *Actinomycete* K-26. The Journal of Antibiotics. 1986;**39**:44-52. DOI: 10.7164/antibiotics.39.44

[98] Hirayama N, Kasai M, Hirata K. Structure and conformation of a novel

inhibitor of angiotensin I converting enzyme—A tripeptide containing phosphonic acid. International Journal of Peptide and Protein Research. 1991;**38**:20-24. DOI: 10.1111/j.1399-3011.1991.tb01404.x

[99] Kido Y, Hamakado T, Anno M, Miyagawa E, Motoki Y, Wakamiya T, et al. Isolation and characterization of I5B2, a new phosphorus containing inhibitor of angiotensin I converting enzyme produced by *Actinomadura* sp. The Journal of Antibiotics. 1984;**37**:965-969. DOI: 10.7164/antibiotics.37.965

[100] Kramer GJ, Mohd A, Schwager LSU, Masuyer G, Acharya KR, Sturrock ED, et al. Interkingdom pharmacology of angiotensin-I converting enzyme inhibitor phosphonates produced by *Actinomycetes*. ACS Medicinal Chemistry Letters. 2014;**5**:346-351. DOI: 10.1021/ml4004588

[101] Masuyer G, Cozier GE, Kramer GJ, Bachmann BO, Acharya KR. Crystal structure of a peptidyl-dipeptidase K-26-DCP from *Actinomycetein* complex with its natural inhibitor. The FEBS Journal. 2016;**283**:4357-4369. DOI: 10.1111/febs.13928

[102] Ntai I, Phelan VV, Bachmann BO. Phosphonopeptide K-26 biosynthetic intermediates in *Astrosporangium hypotensionis*. Chemical Communications. 2006:4518-4520. DOI: 10.1039/b611768f

Section 3

Phosphates in Biomaterials

Chapter 7

Calcium Phosphate Cements in Tissue Engineering

Manuel Pedro Fernandes Graça and Sílvia Rodrigues Gavinho

Abstract

Calcium phosphate cements (CPCs) consist of a combination of calcium phosphates and a liquid phase, allowing it to fit into the body where it was inserted. Several chemical compositions have been synthesized, promoting specific characteristics to the cements for applications such as bone augmentation and reinforcement and metal implant fixation. The hardening reaction mechanism is at low temperatures and makes it capable of incorporating different drugs and other biological molecules. In addition to the abovementioned advantages, CPCs have excellent bioactivity and osteoconductivity and the ability to form a bone bond. Its function as osteoconductor can be improved by insertion of growth factors. In addition, it is possible to functionalize it with silver ions and use it as a coating of implants, conferring antibacterial properties. In this chapter the physical, mechanical, chemical, and biological properties and the possibility of using these cements as drug carriers or biomolecules will be discussed.

Keywords: calcium phosphates, bone cements, tissue regeneration, drug delivery, osseointegration, antibacterial properties

1. Calcium phosphate cements

Calcium phosphate cements (CPCs) were proposed by Brown and Chow [1] and LeGeros et al. [2] in the 1980s. In 1990, the first CPC was used commercially in the treatment of maxillofacial defects and fractures [3–5].

CPCs consist of a combination of one or more calcium orthophosphate powders in which a liquid phase, usually water or an aqueous solution, is added, allowing it to be set and hardened at the site of the body where it was implanted. This type of cement hardens through a dissolution reaction and a precipitation process, distinguishing itself from other cements that harden through a polymerization reaction. Over time, new compositions have been synthesized promoting specific characteristics to the cements for various applications such as bone augmentation and strengthening [6–14], fixation of metal implants [15, 16], and vertebral fractures [17–19].

The CPCs have the essential advantage of hardening in vivo through a low-temperature reaction. After mixing, the material becomes moldable, and its noninvasive injection represents an important advantage over conventional calcium phosphate ceramics. The fact that this type of cement does not present an exothermic reaction also makes it able to incorporate different drugs and other biological molecules, allowing its application in treatments by drug delivery [20]. In addition to the aforementioned advantages, CPCs have excellent bioactivity and osteoconductivity and an excellent ability to form a bone bond. Its rate of resorption is also a factor to take into account, since, after modifying its structure, it is possible to modify that rate. Calcium phosphate cements also have disadvantages, namely, their low mechanical performance, which limits their application in bearing situations. Its intrinsic porosity also leads to this material presenting less strength compared to calcium phosphate ceramics [21, 22].

The cements tend to dissolve in order to achieve a stable and less soluble phase, the dissolution being controlled by the pH of the medium. The cements based on hydroxyapatite or brushite are the only final reaction products because they are the most stable at pH > 4.2 and pH < 4.2, respectively, even though there are a large number of formulations. In addition, and because these materials are intended to be used as bone substitutes, it is important to take into account that the values of the compressive strength of the cortical bone vary between 90 and 209 MPa, [23, 24] and the spongy bone varies between 1.5 and 45 MPa. [25]. As reference values, the compressive strength of apatite cements usually ranges from 20 to 50 MPa [26–32]. Brushite CPCs are generally weaker than apatite CPCs being around 25 MPa [33].

In summary, the main characteristics of this type of cement are presented in **Table 1**.

1.1 Physical and mechanical properties

The mechanical properties are the main properties to take into account when developing a biomaterial to apply surgically. The microstructural characteristics (porosity, quantity, size, morphology, and distribution of the crystals formed) of a biomaterial are the determining factor to define the mechanical properties. These characteristics are controlled by the synthesis process and its intrinsic parameters. In addition, all factors, such as chemical composition of cement, relative proportions of reactants, powder or liquid additives acting as accelerators or retarders, particle size, liquid-powder ratio (L/P ratio) (**Figure 1**), applied pressure during synthesis, and aging conditions, will affect its mechanical properties [36].

Unlike bioceramics, which require sintering at high temperatures, CPCs are formed through a dissolution-precipitation process at room or body temperature. During this process, a crystalline matrix is formed in which with the passage of time and matrix it becomes increasingly dense until reaching the maximum mechanical properties [36].

Normally, microporosity ranges between 30 and 55% and is dependent on the L/P ratio; the higher this ratio is, the greater the microporosity [37].

Calcium phosphate properties	
Material type	Ceramic
Liquid phase	Water or aqueous solutions
Powder component	Calcium phosphate powders
Setting reaction mechanism	Dissolutions and precipitation reaction
Reaction products	Calcium phosphates, usually hydroxyapatite or brushite (37°C)
Stability	Resorbable (low or high resorption rate depending on composition and microstructure)
Bioactivity	Bioactive
Applications	Bone regeneration; non-load-bearing applications

Table 1.
The nature and properties of calcium phosphate bone cements (adapted from [34]).

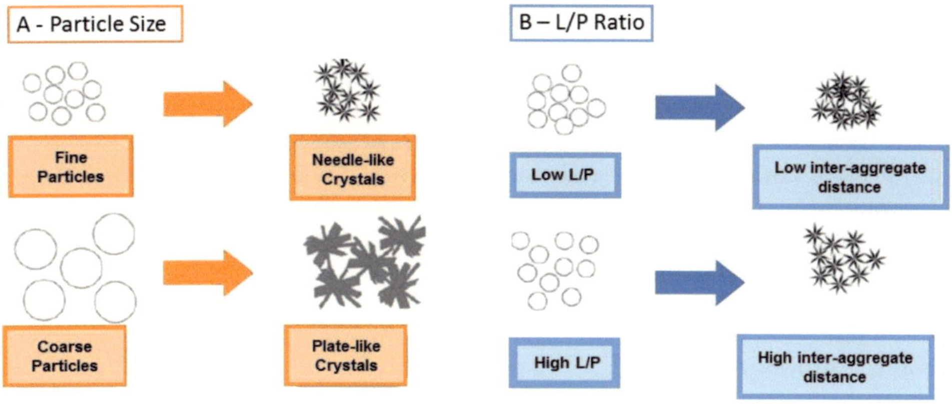

Figure 1.
Microstructure and porosity of CPCs according to particle size and L/P ratio. (A) When the particle size is small, there is an increase in the specific surface area, resulting in the degree of supersaturation. This phenomenon favors the nucleation of crystals and leads to the formation of large quantities of small needle-like crystals. When the particles are larger, the formation of larger plate crystals occurs. Small pores are formed in the cement where small particles are used. (B) Porosity and pore size distribution also vary with the L/P ratio. When the L/P ratio is low, the space between the particles in the mixture decreases, leading to a more compact structure of crystal agglomerates. In contrast, when the L/P ratio increases, the total porosity of the cement increases, and the larger pores are formed due to the increased separation between the aggregates (adapted from [35]).

Although porosity is a disadvantage to be applied in load situations, porosity is sought to improve the resorbability of the material and the extent of bioactivity, increasing the surface area available for reaction. The presence of a certain level of porosity makes this material a good carrier for controlled drug systems [34].

It is possible to promote the creation of macropores in CPCs through two techniques. The leaching of porogen after the adjustment creates macropores; however, it is necessary to add large amounts of porogen, which may compromise the injection process [37–47]. The formation of the pores can also be achieved by the formation of gas foam prior to fixation, but the release of the gas after the introduction of the implant may be harmful to the organism [46, 48–52]. In order to overcome the drawbacks of the techniques mentioned above for obtaining macroporous structures of CPCs, Ginebra et al. proposed the use of self-adjusting injectable macroporous foams composed of a protein-based foaming agent and CPC paste [52].

Regardless of the composition of CPC (apatite or brushite), the strength decreases globally with increased porosity, which is a common occurrence in several materials mainly in porous materials used in bone replacement [36]. In addition to the pore fraction effect, pore size also significantly affects the strength of CPCs. Bai et al. [53] found that the compressive strength is inversely proportional to the size of the macropores through a study in materials with equivalent total porosity but with different sizes of macropores. Through Griffith's classical theory [36], which relates strength to the critical size of the fault, macropores can be considered as failures, thus reducing strength. In addition to the quantity and size of the pores, the characteristics of the crystals (quantity, size, morphology, and distribution) also influence the strength of these cements. The growth of the crystals depends on the kinetics of the dissolution-precipitation reaction of the cement, being controlled by many factors. The smaller the particle size of the starting materials, the faster the material will be converted to apatite, and the crystals formed will not have time to grow. The small size of these crystals will lead to a more dense and crystalline organization, increasing the strength of the cement [54].

One of the conditions that influence the kinetics of apatite formation is aging. The transformation of the initial reactants into apatite becomes faster at higher temperatures, thus making the structure more homogeneous and denser, making it more resistant. In contrast, high temperatures also cause the development of precipitated apatite crystals more rapidly, resulting in larger crystals, which will negatively influence the resistance [55].

The fixation kinetics may be influenced by the presence of accelerating or retarding substances which are added to the mixture and is an important factor in determining the strength of the structure.

Bermudez et al. [56] and Yang et al. [57] found that by adding certain amounts of apatite, the hardening time of the CPCs is lower, and the compressive strength increases considerably. α-Hydroxylic acids (citric acid or glycolic acid) and their salts (sodium citrate) are also used as retarders. The addition of the same allows to mix and to process more easily, reducing the L/P ratio associated with a decrease of the porosity, improving the strength [58–64]. However, it is necessary to take into account an optimum concentration of these additives since the excess may lead to the opposite effect and decrease the force [63]. In summary, the mechanical properties of the CPCs, and in particular the resistance, depend strongly on the microstructure, which is related to the synthesis process, chemical composition, powder or liquid additives acting as accelerators or retarders, particle size, L/P ratio, and aging conditions. In addition, it has been found that crystalline structures have, with smaller crystals, become more compact and homogeneous and appear to give better mechanical properties than those with larger crystals.

Strength has been the main property to be studied when evaluating mechanical performance; however, the CPCs applied in bone defects are also subject to cyclic loading, and the resistance of CPC to fractures cannot be evaluated by strength alone. In order to adequately evaluate the ability to resist fractures, it is also necessary to take into account fracture toughness that describes the strength of a material containing cracks or notches to resist crack propagation [65, 66].

The toughness of a material depends on its nano-/microstructure and on the possibility of promoting the hardening activation mechanism [67, 68]. Without the activation of significant hardening mechanisms, the fracture toughness of CPCs is very low. Due to the low values of toughness, CPCs are very sensitive to defects and failures. The reliability, that is, the likelihood of failure of brittle materials, is also an important factor when one thinks of applying the cement to load-bearing sites. As previously mentioned, it is possible to improve the hardening mechanism by decreasing porosity, as it is the most damaging factor in mechanical performance. To overcome this problem, it is necessary to decrease the volumetric fraction of the pores in order to achieve a more dense matrix by compacting the cement paste prior to hydration.

Studies have shown that compaction pressure would significantly increase tensile strength [69]. However, when the compaction pressure is above 100 MPa, only a slight decrease in porosity is achieved, and the diametral tensile strength is not substantially improved.

In addition, the use of this method to promote the hardening has the same function of decreasing the L/P ratio, which would influence the workability and injectability of cement pastes, which may exclude the application of this cement in minimally invasive surgery.

In order to overcome this disadvantage, researchers added certain amounts of citric acid to the cement liquid to evaluate its effect on the fixability and fixation properties of apatite cements and found that this addition effectively improves the mechanical properties of the cement. According to this prominent effect of citric acid on strength improvement, Barralet et al. [61] and Gbureck et al. [60] added sodium citrate and compacted the resulting cement slurry, obtaining compressive strengths near the resistance of the cortical bone, that this composition can be used in load-bearing locations. This resistance can also be achieved by other factors, especially in the use of citric acid but without applying external pressure, varying the particle size and distribution of powdered reagents [70].

1.2 Chemical and biological properties

The main chemical reaction occurring in the setting mechanism is similar in all these systems of cements and can be understood by analyzing the behavior of the solubility of the existing compounds in the composition [71–73]. During the fixation reaction, the two mechanisms present are dissolution and reprecipitation [23]. The dissolution is activated by the release of the calcium and phosphate ions from the starting materials, leading to supersaturation in the solution. After the ionic concentration reaches a critical value, the nucleation of the new phase occurs, usually around the powder particles. This new phase develops in line with the dissolution of the reactants [74].

In the dissolution/reprecipitation mechanism, the formation of the precipitates depends on the relative stability of the various calcium phosphate salts in the system. The existence of a precipitate that grows in the form of the crystal agglomerates determines the force that a cement can acquire [74]. The solubility phase diagram predicts this reaction, describing the evolution of the solubility of a compound through the logarithm of the total concentration of calcium (or phosphate) as a function of pH [71–73].

The less stable phase of calcium phosphate tends to dissolve to form a more stable and less soluble phase.

As mentioned above, apatite is the most stable calcium phosphate (less soluble at a pH above 4.2 at room temperature), and brushite is most stable at a pH below 4.2 [71].

In these reactions the amount of water consumed is nonexistent, or almost nil, being necessary only for the reagents to become viable and to allow homogeneity in the solution. For this reason, water becomes one of the main contributions to the development of porosity in cement, and, therefore, CPCs are intrinsically porous materials.

In addition to this material, in situ, the body temperature allows its molding after mixing, to be injectable and therefore to be used as a carrier for biological drugs or molecules [75–77].

In the industry, the biological responses of materials have been increasingly a property to be taken into account, focusing on improving cell and tissue CPC interactions, as well as their applications in bone tissue engineering [78–82]. The improvement of these interactions is one important factor for the application of biomaterials and their commercialization for clinical applications.

Studies evaluating the in vivo behavior of CPCs show high levels of biocompatibility and osteoconductivity, stimulating tissue regeneration [5, 82–92]. Most apatite cements are reabsorbed by cell-mediated mechanisms. The function of the osteoclastic cells in this process is to degrade the materials layer by layer, starting from the surface which is in contact with the bone to the nucleus. The biodegradability of apatite CPCs is slow but higher than that of the synthesized hydroxyapatite. As noted above, the rate of degradation of apatite cements is controlled by precipitated hydroxyapatite (PHA) crystallinity, specific surface area, and matrix porosity.

The cements based on brushite have a higher reabsorption rate than apatites due to their superior stability in the biological environment [89–91]. However, there is a possibility in vivo of brushite cements to be transformed into PHA and thus reduce their total degradation rate. In order to retard this reaction or to avoid magnesium salts have been added [92].

At a biological level, the mechanism of dissolution is mediated by the action of physiological solutions or by cell-mediated processes (phagocytosis) [93].

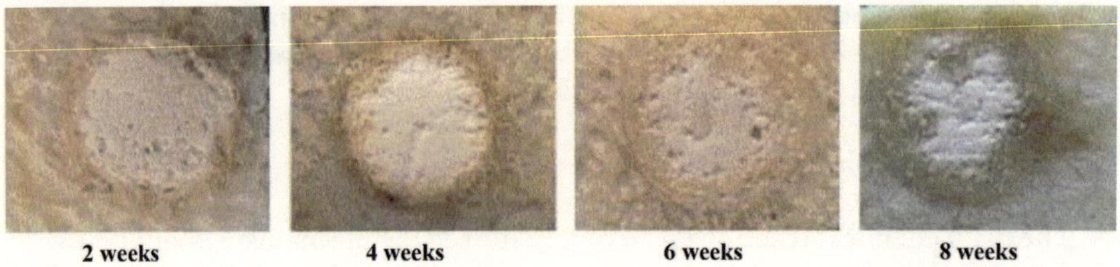

Figure 2.
Image of the drill hole with progressive resorption of the calcium phosphate cement matrix, during 8 weeks [81].

Bone replacement depends on the age, sex, and general metabolic health of the host and the site and volume where it is applied, porosity, crystallinity, chemical composition, particle size, and L/P ratio of the cement. Considering these factors, it can take from 3 to 36 months for the cement to be completely reabsorbed and replaced with bone. However, further studies are needed to confirm the total resorption of the material in order to be applied clinically [94]. Studies have revealed bone development around calcium phosphate cements, demonstrating osteoconductive and osteoinductive characteristics in several cases. It has been shown that within 2 weeks, spicules of living bone with normal bone marrow and gaps in osteocytes can be identified in the cement. After 8 weeks, the cement is almost completely surrounded by new bone. At this stage, no cement reabsorption is typically observed [94]. **Figure 2** shows a progressive resorption of the calcium phosphate cement matrix, with tricalcium phosphate (TCP) granules embedded in a matrix of dicalcium phosphate dihydrate (DCPD) and parallel new bone formation, in a drill hole. After 2 weeks almost the entire surface of the cement was in direct contact with the margins of the bone defect. After 4 weeks, occasional granules of β-TCP and the newly formed bone islets are visible. This area expanded after 6 weeks, involving a progressive reabsorption of the cement matrix and parallel neoformed formation [81].

2. Principal calcium phosphate cements

2.1 Apatite cements

The importance given in the use of apatites in bone replacement is due to the fact that this mineral is the base of the main inorganic part of hard tissues. In fact, nonstoichiometric or calcium-deficient hydroxyapatite (CDHA) is the main mineral phase characteristic of human bones [94]. The CPCs consist of a network of calcium phosphate crystals, with chemical composition and crystal size that can be modified to approximate the biological hydroxyapatite that exists in the living bone [95, 96].

In this regard, it is necessary to clarify that even though the stoichiometric hydroxyapatite has a fixed composition, the apatite structure may exist in a variety of compositions. CDHA comprises in its composition the possibility of varying amounts of calcium, where it is possible to present a completely deficient structure based on this base element. The composition may be expressed as $Ca10 - x(HPO4)(PO4)6 - x(OH)2 - x$, where x ranges from 0 to 1, 0 for stoichiometric hydroxyapatite and 1 for hydroxyapatite totally deficient in calcium. Biological apatite is deficient in calcium containing various ionic substitutions such as Na^+, K^+, Mg^{2+}, F^-, and Cl^- [34].

Apatite cements may form PHA or CDHA through a precipitation reaction. The synthesis of these cements allows the incorporation of different ions in their composition, depending on the initial compounds. The formation of hydroxyapatite that occurs in the cement is compared to the process of formation of new bone and is also seen as a biomimetic process, because it occurs at body temperature and physiological environment. This may explain the fact that the hydroxyapatite formed in the reactions of calcium phosphate cements is much more similar to biological apatites than the ceramic hydroxyapatite resulting from high-temperature sintering processes [34].

The CPCs lead to the formation of PHA or CDHA and can be divided into three systems (single compound, two compounds, more than two compounds), taking into account the number and type of calcium phosphates used in the synthesis [97].

Monocomponent CPCs are those having a single calcium phosphate reagent that hydrolyzes to form PHA or CDHA. Taking into account that at pH 4.2 the hydroxyapatite is less soluble, any other calcium phosphate present will dissolve, and the PHA will tend to precipitate. However, when the formation of PHA occurs, from the hydrolysis of calcium phosphate, the reaction mechanism becomes very slow, due to a decrease in the level of supersaturation, as the reaction proceeds [13]. In this in which only one compound is present, no release of any acid or no base occurs due to the Ca and P ratio being maintained [34, 35].

The second type of cement that may exist is composed of two calcium phosphates, one acid and the other basic, where they adjust after an acid–base reaction. The most commonly used compound is usually tetracalcium phosphate (TTCP), as it is the only calcium phosphate with a Ca/P ratio higher than PHA. Therefore, TTCP can be combined with one or more calcium phosphates with lower Ca/P ratios to obtain PHA or CDHA, avoiding the formation of acids or bases as final products. The combinations that have been more studied seek to produce cements that adjust to the body temperature in a range of pH around the neutral [34].

The third possible system consists of more than two compounds, including calcium phosphates and other salts. For example, a cement proposed by Norian Corporation [5] is used where calcium phosphates with a Ca/P ratio lower than PHA and CaCO3 are added as an additional source of calcium. The initial configuration process involves the formation of DCPD, later forming dahllite, a carbonated hydroxyapatite similar to the bone mineral [5].

Apatitic CPCs appear as a viscous, easily moldable material; however, their injection is difficult. **Figure 3** shows the microstructure of an apatitic cement after setting. The setting time can also be reduced by means of additives such as with the introduction of PHA particles. These changes in the composition may lead to an

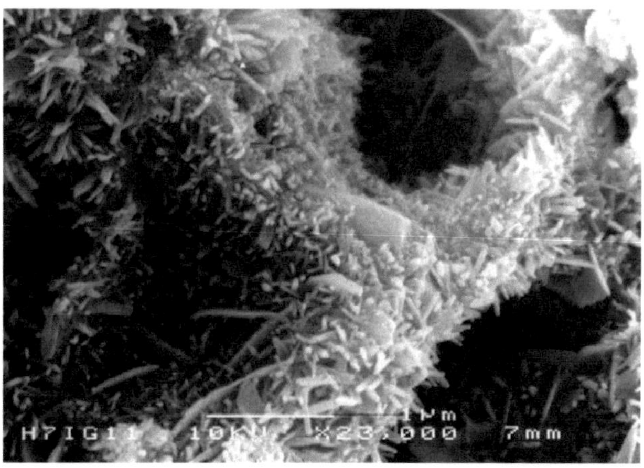

Figure 3.
Microstructure of an apatitic calcium phosphate cement after setting, showing the micro−/nanosize pore structure formed by the entanglement of the precipitated crystals [71].

adjustment time in the range of about 15 minutes. When hardening of the cement paste occurs too fast, the hardened cement must be milled to render it viscous again. Subsequently, the paste hardens due to the precipitation of PHA.

After implantation, the mechanical properties can be altered. Investigations indicate that the mechanical properties of apatite CPC tend to increase, unlike brushite cement, which initially decrease and increase when the bone develops [98, 99].

2.2 Brushite cements

Brushite (DCPD) is an acidic calcium phosphate that has been found in some physiological sites, for example, in bones [100]. Unlike hydroxyapatite, brushite is metastable under physiological conditions [101] and for this reason reabsorbed much faster than CPC apatite; however, there are studies that conclude that DCPD in vivo tends to convert to PHA [26]. Some CPCs were designed to provide brushite as the final product.

Several combinations of compounds have been proposed for the formation of brushite cements; most are β-tricalcium phosphate (β-TCP) and an acid component, namely, monocalcium phosphate monohydrate (MCPM) or phosphoric acid [102–104].

The reaction leading to the formation of brushite CPCs is an acid–base reaction. The brushite paste is acidic during sedimentation because brushite can only precipitate at a pH value below ~6 [105]. The pH of the cement paste tends to change slowly toward equilibrium pH [106]. If the slurry contains an excess of basic phase, the pH tends to equilibrate by crossing the solubility of the base phase with that of the DCPD. The time of stabilization of brushite CPC depends greatly on the solubility of the basic phase: the higher the solubility of the basic phase, the faster the defined time. For example, hydroxyapatite (HA) + MCPM blends have an adjustment time of several minutes. The β-TCP + MCPM mixtures have an adjustment time of 30–60 seconds [107, 108]. However, compounds that inhibit the development of DCPD crystals can be added, increasing the settling time of the β-TCP + MCPM mixtures [109]. The brushite CPC can initially be very liquid and still be defined within a short period of time, unlike the apatite CPC. The brushite CPC is slightly weaker (tensile strength of 10 MPa [110] and compressive strength of 60 MPa) than the apatite CPC (tensile strength of 16 MPa [111] and compressive strength of 83 MPa [112]). The mechanical properties of apatite CPC increase [98], whereas those of brushite CPC decrease [99] This latter phenomenon is attributed to the greater solubility of DCPD in relation to that of PHA [113]. After a few weeks of implantation, the mechanical properties of brushite CPC are promoted by bone growth [99]. Although brushite CPC exhibits biocompatible properties, inflammatory reactions have been reported with the excessive addition of brushite CPC [114]. Investigations indicate that these reactions are due to the transformation of DCPD into PHA [115]. This reaction releases large amounts of acid. The transformation of DCPD into PHA can be avoided with the addition of magnesium ions to the cement [116]. Unlike apatite CPC, brushite CPC cannot be reabsorbed exclusively by osteoclastic activity but also by simple dissolution. Therefore, brushite CPCs degrade at a faster rate than the apatite CPC.

Although brushite demonstrates a higher solubility rate than the other calcium phosphate phases, it is a precursor of the most stable HA phase [117–120]. For this reason, DCPD coatings as an initial step to obtaining HA have been widely used. The synthesis of HA through precipitation mechanisms results in compacted crystals but with sizes difficult to control. Using DCPD as a precursor becomes favorable since it is possible to modify the crystal size of the DCPD through homogeneous precipitation and can be converted directly into HA [119]. In environments with a pH > 6–7, brushite becomes unstable and becomes the most favorable HA phase [121, 122].

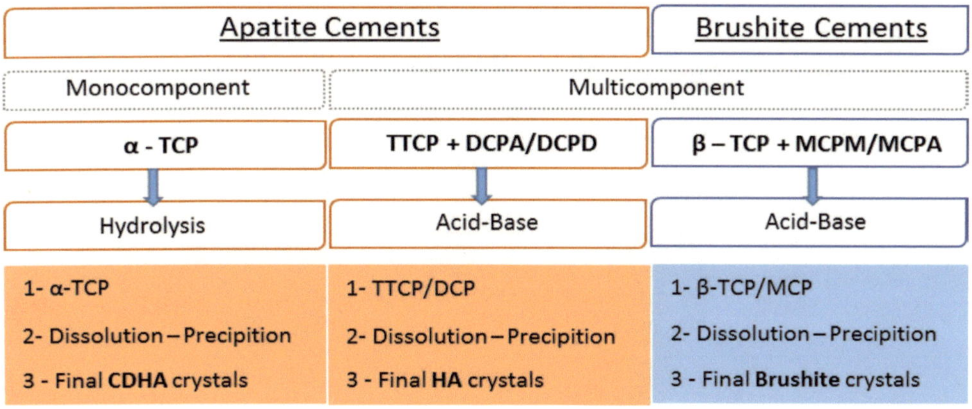

Figure 4.
Classification of calcium phosphate cements, with examples of the most common formulations. From top to bottom, the cements are classified by the type of end product (apatite or brushite), a number of components in the solid phase (single or multiple), type of setting reaction (hydrolysis or acid–base reaction), setting mechanism, and microstructure evolution during setting (adapted from [35]).

The fact that DCPD is able to be more soluble leads to its use in metal implants as a means of increasing the amount of calcium and phosphate ions available in the surrounding tissue of the implant to promote increased osseointegration [118].

The biocompatibility of DCPD as a coating has been demonstrated in several cell lines as, for example, in pre-osteoblastic macrophages [123, 124] and fibroblastic cells [125]. The biocompatibility of DCPD has also been demonstrated when used at a cranial defect site in sheep [126], and the formation of new bone was observed in the absence of inflammation [81]. A clinical study in humans in 2010 effectively used a brushite cement for the repair and increase of pterionic craniotomies, with no inflammation occurring [127].

Figure 4 summarizes how the CPCs are classified by type and number of initial reagents as well as their final product.

3. Calcium phosphate cement applications

3.1 Drug delivery

The main requirements for a substrate to have potential as a drug carrier are to have the ability to incorporate it, to retain it at a specific site, and to distribute it progressively over time in surrounding tissues. In addition, it is beneficial that the material is injectable and biodegradable [77].

Calcium phosphate cements, in addition to allowing hardening at room or body temperature, allow the insertion of various components due to their intrinsic porosity. It is possible to incorporate drugs, biologically active molecules, or even cells without their functions being altered by the effect of temperature or even losing their activity during the procedure (**Figure 5**). This change in the CPCs offers new properties in addition to the osteoconductive characteristic, namely, to increase its capacity for bone regeneration or to support in disorders or pathologies, such as bone tumors or osteoporosis [35].

It is necessary to take into account that the performance of the drug delivery depends on the structural characteristics such as the specific surface area, permeability, matrix degradation rate, drug solubility, or the interaction itself between the matrix and the inserted drug.

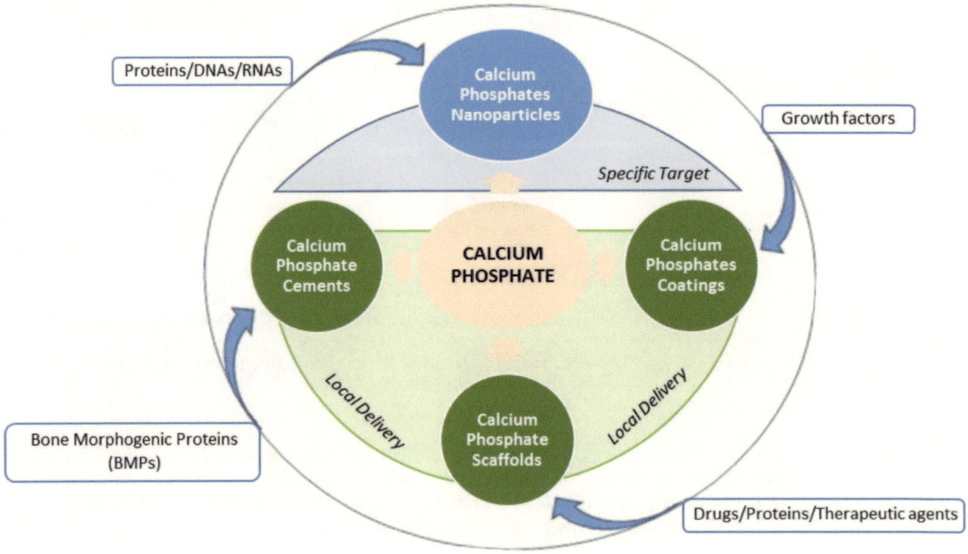

Figure 5.
Scheme about the possible applications of calcium phosphates in the biomolecules and drug delivery (adapted from [128]).

The interaction between the drug and the cement matrix is defined by the drug insertion procedure. Usually, the drugs are added as a powder to the solid phase or dissolved in the liquid part. The method that allows better homogenization of the drug in the matrix is in the liquid phase [35].

Another method for inserting the drug into the matrix is by impregnating solid beads or granules from the cement with a drug solution. This method continues to present benefits compared to other conventional methods even with its impaired injection process. The advantages associated with this method are due to the fact that the consolidation of the material (dissolution-precipitation reaction) leads to the production of hydrated matrices with large specific surface areas that allow the loading of drugs and their release mechanism to be favorable [35].

The two drug encapsulation processes differ greatly in the structural properties of the matrix. In one method the drug is incorporated in the initial phase together with the cement reactants while the matrix is still evolving. Thus, the hardening can last for hours or even days until the suspension of particles evolves into a network of interlaced crystals. In the other case, if the drug is added to the preconceived cement, the matrix structure will always be stable throughout the release process, in addition to possible degradation, and therefore the results cannot be extrapolated to the previous situation. This highlights the need for the studies to be carried out, taking into account the actual conditions of application. The alternative method is to incorporate the drug into polymeric microspheres prior to mixing in the cement. This procedure has two advantages over the other methods presented. It is possible to modify the release kinetics of the drug, and the degradation of the microspheres generates an array capable of being more easily reabsorbed and remodeled [35].

The incorporation of drugs can influence the entire mechanism of action of the cement and compromise its purpose, changing the fit kinetics, rheological characteristics, and microstructural development. For example, molecules that interact with calcium or phosphate ions promote a coprecipitation during the fit or form complexes with Ca^{2+}, promoting a delay in precipitation and modifying viscosity, set time, and cement properties [72, 129, 130].

In addition to the influence that the drug has on the cement matrix, it is also necessary to take into account the influence of the cement on the stability of the drug or bioactive molecule. Due to the dissolution-precipitation process, there is a change in the surrounding pH as well as the change in ionic concentrations, which may influence the functionality of the drug and its release.

Thus, it is beneficial to study the release of drugs introduced from already synthesized cements.

Recently, studies have been developed related to the incorporation of antibiotics, as a preventive method of infections resulting from surgeries or as treatment of bone infections. In addition to antibiotics, studies with anti-inflammatories, antitumor drugs, or hormones have also been disclosed. In another aspect, the incorporation of factors that stimulate bone regeneration, such as bone morphogenetic proteins (BMP) or transforming growth factors-β (TGF-β), has been studied [77, 131].

3.2 Growth factor addition

Growth factors are a large group of proteins that interact at the cellular level [35]. The major families of these proteins are the transforming growth factor-beta superfamily responsible for promoting bone regeneration. The BMPs are part of the TGF-β superfamily and have been widely used in bone regeneration. It is known to play a role as an activating agent in the various biological phenomena responsible for bone formation and therefore can accelerate bone growth. These BMPs stand out from the other growth factors of the large TGF-βSF group because they are osteoinductive. That is, the BMPs act at the level of cell differentiation transforming the pluripotential cells into bone-forming cells, aiding in bone formation outside the bone tissue [77].

These proteins have been produced at an industrial level with a high level of purity; however, it is necessary that their administration is controlled and with adequate therapeutic levels as well as adapted to the tissue targets. In fact, it is known that the injection of such substances alone cannot induce the formation and regeneration of tissues since the protein diffuses very rapidly from the site of implantation. Thus, the CPCs present themselves as good substrates and carriers for these bioactive molecules, also improving their function as osteoconduction [77].

Studies have shown that the superfamily of growth factors stimulates osteoblast proliferation and collagen synthesis in vitro [132] and may increase the size of the cortical bone when applied near the periosteum in vivo [133].

This improvement in bone growth, when applied to cement with these molecules, is due to the adsorption of large doses of rhTGF-β_1 on the surface of the material [134, 135].

Despite this accumulation of the growth factor to the surface, there is a homogenous distribution throughout the cement mass, increasing the time of the release of the growth factors while the degradation of the matrix occurs.

Blom et al. showed that the addition of a human recombinant TGF-β_1 (rhTGF-β_1) to a CPC in the adjustment phase stimulated the differentiation of preosteoblastic cells using primary mouse bone cells in vitro [136].

In contrast to the kinetics of drug release, the release of these factors becomes much slower [137].

It has been determined that in the first days, the release rate of the components is higher because the initial release is only of the material present in the surface layer that is in contact with the medium. This increase in the rhTGF-β_1 release was confirmed when the area in contact with the medium occurred by fragmentation. The same phenomenon was observed when BMP-2 human recombinant microspheres were introduced into the CPC. The release of the factor was quite limited due to the possible physical entrapment of the microparticles inside the porous cement. According to the authors, the nanoporosity of CPC not only did not facilitate the release of the protein but could also limit it because of the high binding affinity of the protein by CPC [138].

Haddad et al. [139] investigated the action of implantation of cement loaded with BMP-2 in the bone repair of a critical-sized calvarial vault defect in rabbits. Compared with control, an increase in bone formation was observed at 45% after 12 weeks of implantation.

Other investigations by Seeherman et al. [140, 141] also demonstrated the efficiency of these combined systems (BMP-2/cement). For example, these composites accelerated the filling of a bone defect by 40% after approximately 4 months of implantation, compared to cement without the protein. This study was done in a primate fibula osteotomy [140].

The composite used in the abovementioned study was also used in rabbit bone defects. After 4 weeks of implantation, an acceleration of reabsorption was observed as well as filling of the defect compared to the base cement. This acceleration led to the complete filling of the defect with new bone 8 weeks after the implant.

3.3 Ion addition

To avoid infections resulting from orthopedic surgery, which usually lead to bone loss or subsequent removal of the implant, alternatives such as antibiotic delivery have been used on the site [142–144]. This transport is usually done using poly(methyl methacrylate) (PMMA) or by encapsulating the drug in the CPC matrix. PMMA beads have the drawback that they are not resorbable and require further surgery to remove them and place new antibiotic-loaded spheres if the goal is to prolong the treatment [145–148]. Faced with this drawback, CPCs have been widely studied as degradable materials capable of carrying antibiotics [77, 145, 148–155]. However, there is a risk of creating bacterial resistance due to low doses of release [156–158]. Thus, the use of surface functionalization of biomaterials as well as the coating of implant surfaces with silver ions has been recurring, conferring antibacterial properties [159–163].

The antimicrobial properties of Ag^+ ions have been investigated and studied in the field of biomedical engineering [164]. It has been found that bacteria hardly gain resistance to silver-based products and low concentrations are required to have a bactericidal effect [165].

Therefore, both metallic silver and ionic silver were incorporated in several biomaterials, HA [166–169], and bioactive glasses [164, 170–172]. In both structures, silver has relatively low toxicity to human cells [173–176].

Ewald et al. evaluated the antimicrobial properties of silver-loaded bone cements as well as their osteoconductive and resorbable properties. The reagents used were both α- and β-tricalcium phosphates combined with slightly acidic compounds to form HA or brushite cement. This study revealed that it is possible to synthesize cements with antimicrobial activity with effects comparable to antibiotic treatments. **Figure 6** shows the inhibition of both *S. aureus* and *S. epidermidis* cultured on the surfaces of the silver-doped cements. Ag-brushite exhibits more antibacterial properties than Ag–HA. In addition, in the case of brushite cements, silver ions allow the cements to increase in compressive strength by approximately 30% [177].

Several studies have been carried out using silver-doped calcium phosphate cements, and the results have been satisfactory, demonstrating inhibitory effect against certain bacteria [178].

In addition to silver, other ions have been incorporated into materials composed of calcium phosphate. Doping with Co^{2+} showed proangiogenic effects [179, 180].

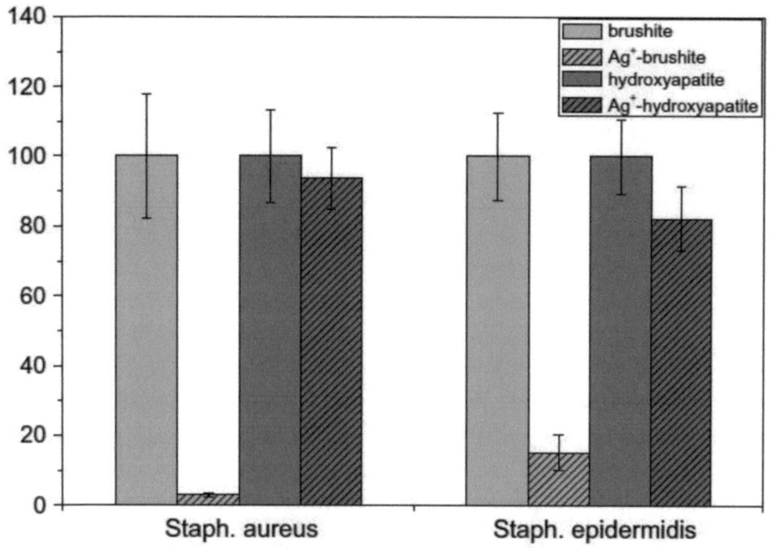

Figure 6.
Bacterial activity of S. aureus *and* S. epidermidis *on cement surfaces (n = 4) determined by the WST-1-test in LB medium [177].*

Modification with Cu^{2+} increased the rate of vascularization [181]. The influence of the introduction of Co^{2+}, Cu^{2+}, and Cr^{3+} on calcium phosphate cement was investigated, evaluating the material properties, proliferation, and osteogenic differentiation of human mesenchymal stem cells in vitro [182]. The Cr^{3+} and Cu^{2+} ions, in this case with less evidence, had positive effects on osteogenic proliferation and differentiation.

Despite all the positive results of this incorporation, the examinations that evaluate the osteogenic capacity in vitro are not enough to estimate the clinical performance of these materials in the bone graft. Since the balance between bone neoformation and material resorption is crucial for successful remodeling, in vitro analysis of osteoclast-mediated degradation of materials is a logical next step in evaluating material remodeling in vivo [183].

However, it is necessary to take into account the dose that is used in the doping process, as investigators have concluded that doses in certain amounts of Cu^{2+} and Co^{2+} can cause cytotoxic reactions to osteoclasts and progenitors of osteoclasts during the initial release of Cu^{2+} and Co^{2+}, respectively. Cements at lower doses are beneficial to bone regeneration since Cu^{2+} at 18 μM completely inhibits reabsorption (but not the formation of osteoclasts), which may be beneficial for patients with osteoporosis and imbalance between bone formation and resorption. The addition of Cr^{3+} to brushite cements increases osteoclastic reabsorption and increases the viability of osteoprogenitor cells compared to cement without the addition of ions [182]. Therefore, Cr^{3+}-doped brushite cements are suggested as a promising new material for application in bone regeneration.

3.4 3D printing of CPCs

The 3D printing technique, or additive manufacture (AM), is based on the addition of layers of powder material producing solid materials with adjustable porosity, through a digital geometric model. This method has been the subject of much research in the medical field such as in the synthesis of the customized scaffold [184]. In addition to the economical and fast production that 3D printing allows, it also allows

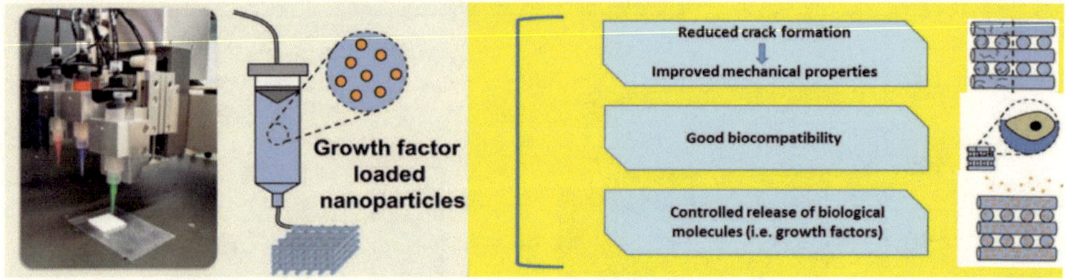

Figure 7.
Scheme of the 3D printing system of the calcium phosphate bone cement scaffolds and its advantages due to the possibility of improving the settings.

the manufacture of pieces with high geometric complexity. All these advantages have made 3D printing very useful in the field of biomedical and tissue engineering due to the ability to replicate the complicated architecture as well as the cellular heterogeneity present in tissues and organs. The bone presents itself as an exquisite structure due to the existence of a complex compound of minerals and an organic matrix. Taking into account this organization, there are several size scales, and it is possible to easily reproduce this structure through 3D printing. In addition, 3D printing is suitable for producing structures derived from medical images, such as CT scans [185, 186].

Recently, the manufacture of the customized scaffolds of the calcium phosphate cements by 3D printing has been described. During printing, highly viscous or pasty materials are dispensed through an adjustable dosing nozzle, resulting in the deposition of CPC layers on a platform with a liquid or air as a plotting medium. After the stabilization of the scaffolds, the water adjustment reaction begins. These stabilization and hardening conditions allow the integration of biological molecules at specific sites [187].

In this way, it is possible to create scaffolds, based on calcium phosphates, more complex with specificities that can promote higher bioactivity or introduce drugs or other therapeutic components, taking into account the patients' needs (**Figure 7**).

In addition, this technique allows the combination of components (i.e., CPC and alginate) to produce structures more resistant to compression and with improved toughness compared to pure CPC supports [188].

4. Conclusions and future perspective

Due to its bioactivity, biocompatibility, osteoconductivity, and osteoinductivity, calcium phosphate cements present an advantageous option in the field of bone tissue engineering, taking into account all the needs that this application demands. In addition, it may be used as scaffolds and transport medium for various biological molecules such as stem cells, drugs, or growth factors. It is also worth mentioning the possibility of producing structures of these cements through 3D printing technology, where it is possible to manufacture intrinsically complex biomimetic structures due to the degree of precision of this technique.

The possibility of building calcium phosphate cements, involving the incorporation of several types of cells, growth factors, molecules, or bioactive glasses, allows favorable results in the vascularization of bone tissues and, consequently, in bone regeneration. This feature, particularly appreciated in large bone regenerations, will allow a considerable increase in the use of these structures in clinical applications. However, more research is needed to consolidate and understand all the information associated with the fundamental mechanisms that promote the development of tissue engineering and regenerative medicine.

The success associated with biomaterials has been underestimated maybe because they have been used clinically for more than 40 years. However, there is still a huge diversity of calcium phosphate-based materials that have not been fully investigated.

Acronyms

AM	manufacture additive
BMP	bone morphogenetic proteins
CDHA	calcium-deficient hydroxyapatite
CPCs	calcium phosphate cements
DCPA	dicalcium phosphate anhydrous
DCPD	dicalcium phosphate dihydrate
HA	hydroxyapatite
L/P ratio	liquid-powder ratio
MCPM	monocalcium phosphate monohydrate
PHA	precipitated hydroxyapatite
PMMA	poly(methyl methacrylate)
TCP	tricalcium phosphate
TGF-β	transforming growth factors-beta
TTCP	tetracalcium phosphate
rhTGF-β_1	human recombinant TGF-β_1

Author details

Manuel Pedro Fernandes Graça* and Sílvia Rodrigues Gavinho
I3N and Physics Department, University of Aveiro, Aveiro, Portugal

*Address all correspondence to: mpfg@ua.pt

References

[1] Brown WE, Chow LC. A new calcium phosphate setting cement. Journal of Dental Research. 1983;**62**:672

[2] LeGeros RZ, Chohayeb A, Shulman A. Apatitic calcium phosphates: Possible dental restorative materials. Journal of Dental Research. 1982;**61**:343

[3] Friedman CD, Costantino PD, Takagi S, Chow LC. BoneSource hydroxyapatite cement: A novel biomaterial for craniofacial skeletal tissue engineering and reconstruction. Journal of Biomedical Materials Research. 1998;**43**:428-432

[4] Kamerer DB, Hirsch BE, Snyderman CH, Costantino P, Friedman CD. Hydroxyapatite cement: A new method for achieving watertight closure in transtemporal surgery. The American Journal of Otology. 1994;**15**:47-49

[5] Constantz BR, Ison IC, Fulmer MT, Poser RD, Smith ST, VanWagoner M, et al. Skeletal repair by in in situ formation of the mineral phase of bone. Science. 1995;**267**:1796-1799

[6] Horstmann WG, Verheyen CCPM, Leemans R. An injectable calcium phosphate cement as a bone-graft substitute in the treatment of displaced lateral tibial plateau fractures. Injury. 2003;**34**:141-144

[7] Strauss EJ, Egol KA. The management of ankle fractures in the elderly. Injury. 2007;**38**:S2-S9

[8] Liverneaux PA. Osteoporotic distal radius curettage—Filling with an injectable calcium phosphate cement. A cadaveric study. European Journal of Orthopaedic Surgery and Traumatology. 2004;**15**:1-6

[9] Welch RD, Zhang H, Bronson DG. Experimental tibial plateau fractures augmented with calcium phosphate cement or autologous bone graft. The Journal of Bone and Joint Surgery. 2003;**85**:222

[10] Aral A, Yalçin S, Karabuda ZC, Anil A, Jansen JA, Mutlu Z. Injectable calcium phosphate cement as a graft material for maxillary sinus augmentation: An experimental pilot study. Clinical Oral Implants Research. 2008;**19**:612-617

[11] Bai B, Jazrawi LM, Kummer FJ, Spivak JM. The use of an injectable, biodegradable calcium phosphate bone substitute for the prophylactic augmentation of osteoporotic vertebrae and the management of vertebral compression fractures. Spine. 1999;**24**:1521-1526

[12] Schildhauer T, Bennett A, Wright T, Lane J, O'Leary P. Intravertebral body reconstruction with an injectable in situ-setting carbonated apatite: Biomechanical evaluation of a minimally invasive technique. Journal of Orthopaedic Research. 1999;**17**:67-72

[13] Maestretti G, Cremer C, Otten P, Jakob RP. Prospective study of standalone balloon kyphoplasty with calcium phosphate cement augmentation in traumatic fractures. European Spine Journal. 2007;**16**:601-610

[14] Libicher M, Hillmeier J, Liegibel U, Sommer U, Pyerin W, Vetter M, et al. Osseous integration of calcium phosphate in osteoporotic vertebral fractures after kyphoplasty: Initial results from a clinical and experimental pilot study. Osteoporosis International. 2006;**17**:1208-1215

[15] Mermelstein LE, Chow LC, Friedman CD, Crisco JJ. The reinforcement of cancellous bone screws with calcium phosphate cement. Journal of Orthopaedic Trauma. 1996;**10**:15-20

[16] Ooms E, Wolke J, Van der Waerden J, Jansen J. Use of injectable calcium-phosphate cement for the fixation of titanium implants: An experimental study in goats. Journal of Biomedical Materials Research. Part B, Applied Biomaterials. 2003;**66**:447-456

[17] Takemasa R, Kiyasu K, Tani T, Inoue S. Validity of calcium phosphate cement vertebroplasty for vertebral non-union after osteoporotic fracture with middle column involvement. The Spine Journal. 2007;7:148S

[18] Tomita S, Kin A, Yazu M, Abe M. Biomechanical evaluation of kyphoplasty and vertebroplasty with calcium phosphate cement in a simulated osteoporotic compression fracture. Journal of Orthopaedic Science. 2003;**8**:192-197

[19] Lewis G. Injectable bone cements for use in vertebroplasty and kyphoplasty: State-of-the-art review. Journal of Biomedical Materials Research. Part B, Applied Biomaterials. 2006;**76**:456-468

[20] Ginebra MP, Traykova T, Planell JA. Calcium phosphate cements: Competitive drug carriers for the musculoskeletal system? Biomaterials. 2006;**27**:2171-2177

[21] Ginebra MP. Cements as bone repair materials. In: Planell JA, editor. Bone Repair Biomaterials. Cambridge, UK: Woodhead Publishing Limited; 2009. pp. 271-308

[22] Canal C, Ginebra MP. Fibre-reinforced calcium phosphate cements: A review. The Journal of the Mechanical Behavior of Biomedical Materials. 2011;**4**:1658-1671

[23] Ontañón M, Aparicio C, Ginebra MP, Planell JA. Structure and mechanical properties of bone. In: Elices M, editor. Structural Biological Materials, Pergamon Material Series. Oxford, UK: Elsevier Science Ltd; 2000. pp. 31-71

[24] Burstein AH, Reilly DT, Martens M. Aging of bone tissue: Mechanical properties. The Journal of Bone and Joint Surgery. 1976;**58A**:82-86

[25] Carter DR, Hayes WC. The compressive behavior of bone as a two-phase porous structure. Clinical Orthopaedics and Related Research. 1977;**59A**:954-962

[26] Constantz BR, Barr BM, Ison IC, Fulmer MT, Baker J, Mckinney LA, et al. Histological, chemical, and crystallographic analysis of four calcium phosphate cements in different rabbit osseous sites. Journal of Biomedical Materials Research Part B: Applied Biomaterials. 1998;**43**:451-461

[27] Ginebra MP, Fernández E, De Maeyer EAP, Verbeeck RMH, Boltong MG, Ginebra J, et al. Setting reaction and hardening of an apatitic calcium phosphate cement. Journal of Dental Research. 1997;**76**(4):905-912

[28] Ginebra MP, Driessens FCM, Planell JA. Effect of the particle size on the micro and nanostructural features of a calcium phosphate cement: A kinetic analysis. Biomaterials. 2004;**25**:3453-3462

[29] Driessens FCM, Planell JA, Boltong MG, Khairoun I, Ginebra MP. Osteotransductive bone cements. Proceedings of the Institution of Mechanical Engineers, Part H: Journal of Engineering in Medicine. 1998;**212**:427-435

[30] Ishikawa K, Miyamoto Y, Kon M, Nagayama M, Asaoka K. Non-decay type fast-setting calcium phosphate cement: composite with sodium alginate. Biomaterials. 1995;**16**:527-532

[31] Khairoun I, Boltong MG, Driessens FC, Planell JA. Effect of

calcium carbonate on clinical compliance of apatitic calcium phosphate bone cement. Journal of Biomedical Materials Research. 1997;**38**(4):356-360

[32] Khairoun I, Driessens FCM, Boltong MG, Planell JA, Wenz R. Addition of cohesion promotors to calcium phosphate cements. Biomaterials. 1999;**20**:393-398

[33] Grover LM, Gbureck U, Wright AJ, Tremayne M, Barralet JE. Biologically mediated resorption of brushite cement in vitro. Biomaterials. 2006;**27**:2178-2185

[34] Ginebra MP. Calcium phosphate bone cements. In: Deb S, editor. Ortopedic Bone Cements. Boca Raton, FL: CRC Press; 2008. pp. 206-230

[35] Ginebra MP, Canal C, Espanol M, Pastorino D, Edgar B. Calcium phosphate cements as drug delivery materials. Advanced Drug Delivery Reviews. 2012;**64**:1090-1110

[36] Zhang J, Liu W, Schnitzler V, Tancret F, Bouler JM. Calcium phosphate cements for bone substitution: Chemistry, handling and mechanical properties. Acta Biomaterialia. 2014;**10**: 1035-1049

[37] Zhang JT, Tancret F, Bouler JM. Fabrication and mechanical properties of calcium phosphate cements (CPC) for bone substitution. Materials Science & Engineering. C, Materials for Biological Applications. 2011;**31**:740-747

[38] Guo H, Su JC, Wei J, Kong H, Liu CS. Biocompatibility and osteogenicity of degradable Ca-deficient hydroxyapatite scaffolds from calcium phosphate cement for bone tissue engineering. Acta Biomaterialia. 2009;**5**:268-278

[39] Barralet JE, Grover L, Gaunt T, Wright AJ, Gibson IR. Preparation of macroporous calcium phosphate cement tissue engineering scaffold. Biomaterials. 2002;**23**:3063-3072

[40] Takagi S, Chow LC. Formation of macropores in calcium phosphate cement implants. The Journal of Materials Science: Materials in Medicine. 2001;**12**:135-139

[41] Li M, Liu XY, Liu XD, Ge BF, Chen KM. Creation of macroporous calcium phosphate cements as bone substitutes by using genipin-crosslinked gelatin microspheres. The Journal of Materials Science: Materials in Medicine. 2009;**20**:925-934

[42] Xu H, Quinn JB, Takagi S, Chow LC, Eichmiller FC. Strong and macroporous calcium phosphate cement: Effects of porosity and fiber reinforcement on mechanical properties. Journal of Biomedical Materials Research. 2001;**57**:457-466

[43] Cama G, Barberis F, Botter R, Cirillo P, Capurro M, Quarto R, et al. Preparation and properties of macroporous brushite bone cements. Acta Biomaterialia. 2009;**5**:2161-2168

[44] Félix Lanao RP, Leeuwenburgh SCG, Wolke JGC, Jansen JA. In vitro degradation rate of apatitic calcium phosphate cement with incorporated PLGA microspheres. Acta Biomaterialia. 2011;**7**:3459-3468

[45] Habraken WJEM, Liao HB, Zhang Z, Wolke JGC, Grijpma DW, Mikos AG, et al. In vivo degradation of calcium phosphate cement incorporated into biodegradable microspheres. Acta Biomaterialia. 2010;**6**:2200-2211

[46] Klijn RJ, van den Beucken J, Lanao R, Veldhuis G, Leeuwenburgh SC, Wolke J, et al. Three different strategies to obtain porous calcium phosphate cements: Comparison of performance in a rat skull bone augmentation model. Tissue Engineering. Part A. 2012;**18**: 1171-1182

[47] Qi XP, Ye JD. Mechanical and rheological properties and injectability

of calcium phosphate cement containing poly (lactic-co-glycolic acid) microspheres. Materials Science & Engineering. C, Materials for Biological Applications. 2009;**29**:1901-1906

[48] Almirall A, Larrecq G, Delgado JA, Martinez S, Planell JA, Ginebra MP. Fabrication of low temperature macroporous hydroxyapatite scaffolds by foaming and hydrolysis of an a-TCP paste. Biomaterials. 2004;**25**:3671-3680

[49] Del Real RP, Wolke J, Vallet-Regi M, Jansen JA. A new method to produce macropores in calcium phosphate cements. Biomaterials. 2002;**23**: 3673-3680

[50] Chen WC, Zhou HZ, Tang MH, Weir MD, Bao CY, Xu H. Gas-foaming calcium phosphate cement scaffold encapsulating human umbilical cord stem cells. Tissue Engineering. Part A. 2012;**18**:816-827

[51] Hesaraki S, Moztarzadeh F, Sharifi D. Formation of interconnected macropores in apatitic calcium phosphate bone cement with the use of an effervescent additive. Journal of Biomedical Materials Research. Part A. 2007;**83A**:80-87

[52] Ginebra M, Delgado J, Harr I, Almirall A, Del Valle S, Planell JA. Factors affecting the structure and properties of an injectable self-setting calcium phosphate foam. Journal of Biomedical Materials Research. Part A. 2007;**80A**:351-361

[53] Bai F, Meng GL, Yuan YA, Liu CS, Wang Z, Liu JA. Role of macropore size in the mechanical properties and in vitro degradation of porous calcium phosphate cements. Materials Letters. 2010;**64**:2028-2031

[54] Liu CS, Shao HF, Chen FY, Zheng HY. Effects of the granularity of raw materials on the hydration and hardening process of calcium phosphate cement. Biomaterials. 2003;**24**:4103-4113

[55] TenHuisen KS, Brown PW. Formation of calcium-deficient hydroxyapatite from α-tricalcium phosphate. Biomaterials. 1998;**19**: 2209-2217

[56] Bermudez O, Boltong MG, Driessens FCM, Planell JA. Development of some calcium-phosphate cements from combinations of α-TCP, MCPM and CaO. The Journal of Materials Science: Materials in Medicine. 1994;**5**:160-163

[57] Yang QZ, Troczynski T, Liu DM. Influence of apatite seeds on the synthesis of calcium phosphate cement. Biomaterials. 2002;**23**:2751-2760

[58] Tamimi F, Sheikh Z, Barralet J. Dicalcium phosphate cements: Brushite and monetite. Acta Biomaterialia. 2012;**8**:474-487

[59] Sarda S, Fernandez E, Nilsson M, Balcells M, Planell JA. Kinetic study of citric acid influence on calcium phosphate bone cements as water-reducing agent. Journal of Biomedical Materials Research. 2002;**61**:653-659

[60] Gbureck U, Barralet JE, Spatz K, Grover LM, Thull R. Ionic modification of calcium phosphate cement viscosity. Part I: Hypodermic injection and strength improvement of apatite cement. Biomaterials. 2004;**25**:2187-2195

[61] Barralet JE, Hofmann M, Grover LM, Gbureck U. High-strength apatitic cement by modification with α-hydroxy acid salts. Advanced Materials. 2003;**15**:2091-2094

[62] Barralet JE, Grover LM, Gbureck U. Ionic modification of calcium phosphate cement viscosity. Part II: Hypodermic injection and strength improvement of brushite cement. Biomaterials. 2004;**25**:2197-2203

[63] Qi XP, Ye JD, Wang YJ. Improved injectability and in vitro degradation of a calcium phosphate cement containing poly(lactide-co-glycolide) microspheres. Acta Biomaterialia. 2008;**4**:1837-1845

[64] Marino FT, Torres J, Hamdan M, Rodriguez CR, Cabarcos EL. Advantages of using glycolic acid as a retardant in a brushite forming cement. Journal of Biomedical Materials Research. 2007;**83B**:571-579

[65] Munz D, Fett D. Ceramics: Mechanical Properties, Failure Behavior, Materials Selection. Germany: Springer; 2001

[66] Munz D. What can we learn from r-curve measurements? Journal of the American Ceramic Society. 2007;**90**:1-15

[67] Ritchie RO. The conflicts between strength and toughness. Nature Materials. 2011;**10**:817-822

[68] Launey ME, Ritchie RO. On the fracture toughness of advanced materials. Advanced Materials. 2009;**21**:2103-2110

[69] Chow LC, Hirayama S, Takagi S, Parry E. Diametral tensile strength and compressive strength of a calcium phosphate cement: Effect of applied pressure. Journal of Biomedical Materials Research. 2000;**53**:511-517

[70] Hofmann MP, Mohammed AR, Perrie Y, Gbureck U, Barralet JE. High-strength resorbable brushite bone cement with controlled drug-releasing capabilities. Acta Biomaterialia. 2009;**5**:43-49

[71] Chow LC. Development of self-setting calcium phosphate cements. Journal of the Ceramic Society of Japan (International Edition). 1991;**99**:927-936

[72] Ginebra MP, Rilliard A, Fernández E, Elvira C, San Román J, Planell JA. Mechanical and rheological improvement of a calcium phosphate cement by the addition of a polymeric drug. Journal of Biomedical Materials Research. 2001;**57**:113-118

[73] Driessens FCM, Boltong MG, Bermúdez O, Planell JA, Ginebra MP, Fernández E. Effective formulations for the preparation of calcium phosphate bone cements. Journal of Materials Science. Materials in Medicine. 1994;**5**:164-170

[74] Ginebra MP. Desarrollo y caracterización de un cemento óseo basado en fosfato tricálcico-α para aplicaciones quirúrgicas [PhD thesis]. Barcelona, Spain: Universitat Politècnica de Catalunya; 1996

[75] Chow LC. Calcium phosphate cements: Chemistry, properties, and applications. MRS Proceedings. 1999;**599**:27

[76] Xu HH, Weir MD, Burguera EF, et al. Injectable and macroporous calcium phosphate cement scaffold. Biomaterials. 2006;**27**:4279-4287

[77] Ginebra MP, Traykova T, Planell JA. Calcium phosphate cements as bone drug delivery systems: A review. Journal of Controlled Release. 2006;**113**:102-110

[78] Weir MD, Xu HH. Osteoblastic induction on calcium phosphate cement-chitosan constructs for bone tissue engineering. Journal of Biomedical Materials Research. Part A. 2010;**94**:223-233

[79] Mestres G, Le Van C, Ginebra MP. Silicon-stabilized α-tricalcium phosphate and its use in a calcium phosphate cement: Characterization and cell response. Acta Biomaterialia. 2012;**8**:1169-1179

[80] Lanao RPF, Leeuwenburgh SC, Wolke JG, et al. Bone response to fast-degrading, injectable calcium

phosphate cements containing PLGA microparticles. Biomaterials. 2011;**32**:8839-8847

[81] Theiss F, Apelt D, Brand B, et al. Biocompatibility and resorption of a brushite calcium phosphate cement. Biomaterials. 2005;**26**:4383-4394

[82] Noetzel J, Özer K, Reisshauer B-H, et al. Tissue responses to an experimental calcium phosphate cement and mineral trioxide aggregate as materials for furcation perforation repair: A histological study in dogs. Clinical Oral Investigations. 2006;**10**:77

[83] Kurashina K, Kurita H, Kotani A, Klein CPAT, Groot K. In vivo study of calcium phosphate cements: Implantation of an α-tricalcium phosphate/dicalcium phosphate dibasic/tetracalcium phosphate monoxide cement paste. Biomaterials. 1997;**18**:539-543

[84] Jansen JA, Ruijter JE, Schaeken HG, van der Waerden JPC, Planell JA, Driessens FCM. Evaluation of tricalciumphosphate/hydroxyapatite cement for tooth replacement: An experimental animal study. Journal of Materials Science: Materials in Medicine. 1995;**6**:653-657

[85] FriedmanCD,CostantinoPD,TakagiS, Chow LC. Bonesource hydroxyapatite cement: A novel biomaterial for cranofacial skeletal tissue engineering and reconstruction. Journal of Biomedical Materials Research Part B: Applied Biomaterials. 1998;**43**:428-432

[86] Larsson S, Bauer TW. Use of injectable calcium phosphate cement for fracture fixation: A review. Clinical Orthopaedics and Related Research. 2002;**395**:23-32

[87] Ooms EM, Wolke JGC, van de Heuvel MT, Jeschke B, Jansen JA. Histological evaluation of the bone response to calcium phosphate cement implanted in cortical bone. Biomaterials. 2003;**24**:989-1000

[88] Frankenburg EP, Goldstein SA, Bauer TW, Harris SA, Poser RD. Biomechanical and histological evaluation of a calcium phosphate cement. The Journal of Bone and Joint Surgery. American Volume. 1998;**80**:1112

[89] Apelt D, Theiss F, El-warrak AO, Zlinszky K, Bettschart-Wolfisberger R, Bohner M, et al. In vivo behavior of three different injectable hydraulic calcium phosphate cements. Biomaterials. 2004;**25**:1439-1451

[90] Ohura K, Bohner M, Hardouin P, Lemaitre J, Pasquier G, Flautre B. Resorption of, and bone formation from, new betatricalcium phosphate-monocalcium phosphate cements: An in vivo study. Journal of Biomedical Materials Research. 1996;**30**:193-200

[91] Munting E, Mirtchi AA, Lemaitre J. Bone repair of defects filled with phosphocalcic hydraulic cement: an in vivo study. Journal of Materials Science. Materials in Medicine. 1993;**4**: 337-344

[92] Bohner M, Theiss F, Apelt D, Hirsiger W, Houriet R, Rizzoli G, et al. Compositional changes of a dicalcium phosphate dihydrate cement after implantation in sheep. Biomaterials. 2003;**24**:3463-3474

[93] Lu J, Descamps M, Dejou J, et al. The biodegradation mechanism of calcium phosphate biomaterials in bone. Journal of Biomedical Materials Research. Part A. 2002;**63**:408-412

[94] Dorozhkin SV. Calcium orthophosphate cements and concretes. Materials. 2009;**2**:221-291

[95] Morgan E, Yetkinler D, Constantz B, Dauskardt R. Mechanical properties of carbonated apatite bone mineral

substitute: Strength, fracture and fatigue behavior. Journal of Materials Science. Materials in Medicine. 1997;**8**:559-570

[96] Montufar EB, Traykova T, Schacht E, Ambrosio L, Santin M, Planell JA, et al. Self-hardening calcium deficient hydroxyapatite/gelatine foams for bone regeneration. Journal of Materials Science. Materials in Medicine. 2010;**21**:863-869

[97] Fernández E, Gil FJ, Ginebra MP, Driessens FCM, Planell JA, Best S. Calcium phosphate bone cements for clinical applications. Part II: Precipitate formation during setting reactions. Journal of Materials Science: Materials in Medicine. 1999;**10**:177-183

[98] Miyamoto Y, Ishikawa K, Fukao H, Sawada M, Nagagama M, Kon M, et al. In vivo setting behaviour of fast-setting calcium phosphate cement. Biomaterials. 1995;**16**:855-860

[99] Ikenaga M, Hardouin I, Lemaitre J, Andrianjatovo H, Flautre B. Biomechanical characterization of a biodegradable calcium phosphate hydraulic cement: A comparison with porous biphasic calcium phosphate ceramics. Journal of Biomedical Materials Research. 1998;**40**:139-144

[100] Muenzenberg K, Gebhardt M. Brushite, octocalcium phosphate, and carbonate containing apatite in bone. Clinical Orthopaedics and Related Research. 1973;**90**:271-273

[101] Terjesen T. Bone healing after metal plate fixation and external fixation of the osteotomized rabbit tibia. Acta Orthopaedica Scandinavica. 1984;**55**:69

[102] Lemaitre J, Mirtchi A, Mortier A. Calcium phosphate cements for medical use: State of the art and perspectives of development. Silicates Industriels. 1987;**52**:141-146

[103] Mirtchi AA, Lemaitre J, Terao N. Calcium phosphate cements: Study of the [beta]-tricalcium phosphate–monocalcium phosphate system. Biomaterials. 1989;**10**:475-480

[104] Bajpai P, Fuchs C, McCullum DE. Development of tricalcium phosphate ceramic cements. In: Lemons J, editor. Quantitative Characterization and Performance of Porous Implants for Hard Tissue Applications. Philadelphia, USA: American Society for Testing Materials. 1987. pp. 377-388

[105] Elliott J. Structure and Chemistry of the Apatites and Other Calcium Orthophosphates. Amsterdam: Elsevier; 1994

[106] Bohner M, Van Landuyt P, Merkle H, Lemaitre J. Composition effects on the pH of a hydraulic calcium phosphate cement. The Journal of Materials Science: Materials in Medicine. 1997;**8**:675-681

[107] Lemaitre J, Mirtchi A, Mortier A. Calcium phosphate cements for medical use: state of the art and perspectives of development. Silicates Industriels. 1987;**9-10**:141-6

[108] Bohner M. Proprietes physico-chimiques et osteogeniques d'un biociment hydraulique a base de phosphates de calcium [PhD thesis No. 1171]. Lausanne: Swiss Federal Institute of Technology of Lausanne (EPFL); 1993

[109] Bohner M, Van Landuyt P, Trophardy G, Merkle H, Lemaitre J. Effect of several additives and their admixtures on the physico-chemical properties of a calcium phosphate cement. Journal of Materials Science. Materials in Medicine. 2000;**11**:111-116

[110] Andrianjatovo H, Jose F, Lemaitre J. Effect of b-TCP granulometry on setting time and strength of calcium phosphate hydraulic cements. Journal of Materials Science. Materials in Medicine. 1996;**7**:34-39

[111] Ishikawa K, Takagi S, Chow L, Ishikawa Y, Eanes E, Asaoka K. Behavior of a calcium phosphate cement in simulated blood plasma in vitro. Dental Materials. 1994;**10**:26-32

[112] Driessens F. Chemistry and applied aspects of calcium phosphate bone cements. In: Concepts and Clinical Applications of Ionic Cements. Presented at the 15th European Conference on Biomaterials. ESB 99, Arcachon, France; 1999

[113] Shadanbaz S, Dias GJ. Calcium phosphate coatings on magnesium alloys for biomedical applications: A review. Acta Biomaterialia. 2012;**8**:20-30

[114] Hardouin P, Delecourt C, Blary M, Van Landuyt I, Lemaitre J, Hardouin L. Volume effect on biological properties of a calcium phosphate hydraulic cement: Experimental study in sheep. Bone. 1999;**25**:35-39

[115] Bohner M. Calcium orthophosphates in medicine: From ceramics to calcium phosphate cements. Injury-International Journal of the Care of the Injured. 2000;**31**:S-D37-47

[116] Bohner M, Matter S. Brushite hydraulic cement stabilized with a magnesium salt. PCT application PCT/CH99/00595, Switzerland. 1999

[117] Kumar M, Xie J, Chittur K, Riley C. Transformation of modified brushite to hydroxyapatite in aqueous solution: Effects of potassium substitution. Biomaterials. 1999;**20**:1389-1399

[118] Kumar M, Dasarathy H, Riley C. Electrodeposition of brushite coatings and their transformation to hydroxyapatite in aqueous solutions. Journal of Biomedial Materials Research Part A. 1999;**45**:302-310

[119] Redepenning J, Schlessinger T, Burnham S, Lippiello L, Miyano J. Characterization of electrolytically prepared brushite and hydroxyapatite coatings on orthopedic alloys. Journal of Biomedial Materials Research Part A. 1996;**30**:287-294

[120] Xie J, Riley C, Chittur K. Effect of albumin on brushite transformation to hydroxyapatite. Journal of Biomedical Materials Research. 2001;**57**:357-365

[121] Levinskas G, Neuman W. The solubility of bone mineral. I. Solubility studies of synthetic hydroxylapatite. The Journal of Physical Chemistry. 1955;**59**:164-168

[122] Strates B, Neuman W, Levinskas G. The solubility of bone mineral. II. Precipitation of near-neutral solutions of calcium and phosphate. The Journal of Physical Chemistry. 1957;**61**:279-282

[123] Xia Z, Grover L, Huang Y, Adamopoulos I, Gbureck U, Triffitt J, et al. In vitro biodegradation of three brushite calcium phosphate cements by a macrophage cell-line. Biomaterials. 2006;**27**:4557-4565

[124] Tamimi F, Kumarasami B, Doillon C, Gbureck U, Le Nihouannen D, Cabarcos E, et al. Brushite–collagen composites for bone regeneration. Acta Biomaterialia. 2008;**4**:1315

[125] Klammert U, Reuther T, Jahn C, Kraski B, Kbler A, Gbureck U. Cytocompatibility of brushite and monetite cell culture scaffolds made by three-dimensional powder printing. Acta Biomaterialia. 2009;**5**:727

[126] Kuemmerle J, Oberle A, Oechslin C, Bohner M, Frei C, Boecken I, et al. Assessment of the suitability of a new brushite calcium phosphate cement for cranioplasty—An experimental study in sheep. Journal of Cranio-Maxillo-Facial Surgery. 2005;**33**:37-44

[127] Ji C, Ahn J. Clinical experience of the brushite calcium phosphate cement for the repair and augmentation of surgically induced cranial defects

following the pterional craniotomy. Journal of Korean Neurosurgical Association. 2010;**47**:180

[128] Bose S, Tarafder S. Calcium phosphate ceramic systems in growth factor and drug delivery for bone tissue engineering: A review. Acta Biomaterialia. 2012;**8**:1401-1421

[129] Bigi A, Bracci B, Panzavolta S. Effect of added gelatin on the properties of calcium phosphate cement. Biomaterials. 2004;**25**:2893-2899

[130] Ratier A, Best S, Freche M, Lacout J, Rodriguez F. Behaviour of a calcium phosphate bone cement containing tetracycline hydrochloride or tetracycline complexed with calcium ions. Biomaterials. 2001;**22**:897-901

[131] Nimni ME. Polypeptide growth factors: Targeted delivery systems. Biomaterials. 1997;**18**:1201-1225

[132] Centrella M, Massague J, Canalis E. Human platelet-derived transforming growth factor β stimulates parameters of bone growth in fetal rat calvaria. Endocrinology. 1986;**119**:2306-2312

[133] Noda M, Camilliere JJ. In vivo stimulation of bone formation by transforming growth factor β. Endocrynology. 1989;**124**:2991-2994

[134] Bosch C, Melsen B, Gibbons R, Vargervik K. Human recombinant transforming growth factor beta 1 in healing of calvarial bone defects. Journal of Craniofacial Surgery. 1996;**7**:300-310

[135] Lind M, Overgaard S, Soballe K, Nguyen T, Ongpipattanakul B, Bunger C. Transforming growth factor-beta 1 enhances bone healing to unloaded tricalcium phosphate coated implants: An experimental study in dogs. Journal of Orthopaedic Research. 1996;**14**:343-350

[136] Blom EJ, Klein-Nulend J, Klein CPAT, Kurashina K, van Waas MAJ, Burger EH. Transforming growth factor-β1 incorporated during setting in calcium phosphate cement stimulates bone cell differentiation in vitro. Journal of Biomedical Materials Research. 2000;**50**:67-74

[137] Blom EJ, Klein-Nulend J, Yin L, van Waas MAJ, Burger EH. Transforming growth factor-β1 in calcium phosphate cement stimulates bone regeneration. Journal of Dental Research. 2000;**79**:255

[138] Ruhe PQ, Hedberg EL, Padron NT, Spauwen PH, Jansen JA, Mikos AG. rhBMP-2 release from injectable poly(DL-lactic-co-glycolic acid)/calcium-phosphate cement composites. The Journal of Bone and Joint Surgery. 2003;**85**:75-82

[139] Haddad AJ et al. Closure of rabbit calvarial critical-sized defects using protective composite allogeneic and alloplastic bone substitutes. The Journal of Craniofacial Surgery. 2006;**17**:926-934

[140] Seeherman HJ et al. Recombinant human bone morphogenetic protein-2 delivered in an injectable calcium phosphate paste accelerates osteotomy-site healing in a nonhuman primate model. The Journal of Bone and Joint Surgery. American Volume. 2004;**86-A**: 1961-1972

[141] Seeherman HJ et al. rhBMP-2 delivered in a calcium phosphate cement accelerates bridging of critical-sized defects in rabbit radii. The Journal of Bone and Joint Surgery. American Volume. 2006;**88**:1553-1565

[142] Ruchholtz S, Tager G, Nast-Kolb D. The periprosthetic total hip infection. Unfallchirurg. 2004;**107**:307-317

[143] Harris W, Sledge CB. Total hip and total knee replacement (Part II). The New England Journal of Medicine. 1990;**323**:801-807

[144] Lew DP, Waldvogel FA. Osteomyelitis. The Lancet. 2004;**364**:369-379

[145] Bohner M, Lemaitre J, Van Landuyt P, Zambelli PY, Merkle HP, Gander B. Gentamicin-loaded hydraulic calcium phosphate bone cement as antibiotic delivery system. Journal of Pharmaceutical Sciences. 1997;**86**:565-572

[146] Alt V, Bechert T, Steinrcke P, Wagener M, Seidel P, Dingeldein E, et al. An in vitro assessment of the antibacterial properties and cytotoxicity of nanoparticulate silver bone cement. Biomaterials. 2004;**25**:4383-4391

[147] Itokazu M, Wenyi Y, Aoki T, Ohara A, Kato N. Synthesis of antibiotic-loaded interporous hydroxyapatite blocks by vacuum method and in vitro drug release testing. Biomaterials. 1998;**19**:817-819

[148] Penner MJ, Masri BA, Duncan CP. Elution characteristics of vancomycin and tobramycin combined in acrylic bone-cement. The Journal of Arthroplasty. 1996;**11**:939-944

[149] Dion A, Langman M, Hall G, Filiaggi M. Vancomycin release behaviour from amorphous calcium polyphosphate matrices intended for osteomyelitis treatment. Biomaterials. 2005;**26**:7276-7285

[150] Jiang PJ, Patel S, Gbureck U, Grover LM. A comparison of the efficacy of hydroxyapatite based cements and gels as drug delivery matrices. Key Engineering Materials. 2008;**93**:327-330

[151] Frutos P, Pena E, Frutos G, Barrales-Rienda JM. Release of gentamicin sulphate from modified commercial bone cement. Effect of (2-hydroxyethyl methacrylate) comonomer and poly(N-vinyl-2-pyrrolidone) additive on release mechanism and kinetics. Biomaterials. 2002;**23**:3787-3797

[152] Schnieders J, Gbureck U, Thull R, Kissel T. Controlled release of gentamicin from calcium phosphate–poly(lactic acid-co-glycolic acid) composite bone cement. Biomaterials. 2006;**27**:4239-4249

[153] Anttipoika I, Josefsson G, Konttinen Y, Lidgren L, Santavirta S, Sanzen L. Hip-arthroplasty infection—Current concepts. Acta Orthopaedica Scandinavica. 1990;**61**:163-169

[154] Barralet JE, Aldred S, Wright AJ, Coombes AGA. In vitro behavior of albumin-loaded carbonate hydroxyapatite gel. Journal of Biomedical Materials Research. 2002;**60**:360-367

[155] Gitelis S, Brebach GT. The treatment of chronic osteomyelitis with a biodegradable antibiotic-impregnated implant. Journal of Orthopaedic Surgery. 2002;**10**:53-60

[156] Hendriks JGE, van Horn JR, van der Mei HC, Busscher HJ. Backgrounds of antibiotic-loaded bone cement and prosthesis-related infection. Biomaterials. 2004;**25**:545-556

[157] Pitto RP, Spika IA. Antibiotic-loaded bone cement spacers in two-stage management of infected total knee arthroplasty, International Orthopaedics. 2004;**28**:129-133

[158] Durbhakula SM, Czajka J, Fuchs MD, Uhl RL. Spacer endoprosthesis for the treatment of infected total hip arthroplasty. The Journal of Arthroplasty. 2004;**19**:760-767

[159] Ewald A, Gluckermann SK, Thull R, Gbureck U. Antimicrobial titanium/silver PVD coatings on titanium, BioMedical Engineering Online. 2006;**5**:22

[160] Yorganci K, Krepel C, Weigelt JA, Edmiston CE. Activity of antibacterial impregnated central venous catheters

against Klebsiella pneumonia, Intensive Care Medicine. 2002;**28**:438-442

[161] Davenport K, Keeley FX. Evidence for the use of silver-alloy-coated urethral catheters. The Journal of Hospital Infection. 2005;**60**:298-303

[162] Cook G, Costerton JW, Darouiche RO. Direct confocal microscopy studies of the bacterial colonization in vitro of a silver-coated heart valve sewing cuff. The International Journal of Antimicrobial Agents. 2000;**13**:169-173

[163] Boswald M, Lugauer S, Regenfus A, Braun GG, Martus P, Geis C, et al. Reduced rates of catheter-associated infection by use of a new silver-impregnated central venous catheter. Infection. 1999;**27**:56-60

[164] Blaker JJ, Nazhat SN, Boccaccini AR. Development and characterisation of silver-doped bioactive glass-coated sutures for tissue engineering and wound healing applications. Biomaterials. 2003;**25**:1319-1329

[165] Clement JL, Jarrett PS. Anti-bacterial silver. Metal-Based Drugs. 1994;**1**:467-482

[166] Mahabole M, Aiyer R, Ramakrishna C, Sreedhar B, Khairnar R. Synthesis, characterization and gas sensing property of hydroxyapatite ceramic. The Bulletin of Materials Science. 2005;**28**:535-545

[167] Kim TN, Feng QL, Kim JO, Wu J, Wang H, Chen GC, et al. Antimicrobial effects of metal ions (Ag^{+}, Cu^{2+}, Zn^{2+}) in hydroxyapatite. Journal of Materials Science. Materials in Medicine. 1998;**9**: 129-134

[168] Chen W, Oh S, Ong AP, Oh N, Liu Y, Courtney HS, et al. Antibacterial and osteogenic properties of silver-containing hydroxyapatite coatings produced using a sol gel process. Journal of Biomedical Materials Research. Part A. 2007;**82**:899-906

[169] Chen W, Liu Y, Courtney HS, Bettenga M, Agrawal CM, Bumgardner JD, et al. In vitro anti-bacterial and biological properties of magnetron co-sputtered silver-containing hydroxyapatite coating. Biomaterials. 2006;**27**:5512-5517

[170] Clupper DC, Hench LL. Bioactive response of Ag-doped tape cast bioglass-45S5 following heat treatment. Journal of Materials Science. Materials in Medicine. 2001;**12**:917-921

[171] Sharma VK, Yngard RA, Lin Y. Silver nanoparticles: green synthesis and their antimicrobial activities. Advances in Colloid and Interface Science. 2009;**145**:83-96

[172] Ciceo Lucacel R, Hulpus AO, Simon V, Ardelean I. Structural characterization of phosphate glasses doped with silver. Journal of Non-Crystalline Solids. 2009;**355**:425-429

[173] Hollinger MA. Toxicological aspects of topical silver pharmaceuticals. Critical Reviews in Toxicology. 1996;**26**:255-260

[174] Faust RA. Toxicity Summary for Silver, Prepared for the Oak Ridge Reservation Environmental Restoration Programme. Oak Ridge National Laboratory, US Department of Energy; 1992. Available from: http://rais.ornl.gov/tox/profiles/silver_c_V1.html [Accessed: 19 August 2019]

[175] Berger TJ, Spadaro JA, Chapin SE, Becher RO. Electrically generated silver ions: quantitative effects on bacterial and mammalian cells. Antimicrobial Agents and Chemotherapy. 1976;**9**:357-358

[176] Williams RL, Doherty PJ, Vince DG, Grashoff GJ, Williams DF. The biocompatibility of silver. Critical

Reviews in Biocompatibility. 1989;**5**:221-223

[177] Ewald A, Hösel D, Patel S, Grover LM, Jake E, Barralet JE, et al. Silver-doped calcium phosphate cements with antimicrobial activity. Acta Biomaterialia. 2011;**7**:4064-4070

[178] Rau JV, Fosca M, Graziani V, Egorov AA, Zobkov V, Fedotov A, et al. Silver-doped calcium phosphate bone cements with antibacterial properties. Journal of Functional Biomaterials. 2016;**7**:10

[179] Kulanthaivel S, Roy B, Agarwal T, Giri S, Pramanik K, Pal K, et al. Cobalt doped proangiogenic hydroxyapatite for bone tissue engineering application. Materials Science & Engineering. C, Materials for Biological Applications. 2016;**58**:648-658

[180] Birgani ZT, Gharraee N, Malhotra A, van Blitterswijk CA, Habibovic P. Combinatorial incorporation of fluoride and cobalt ions into calcium phosphates to stimulate osteogenesis and angiogenesis. Biomedical Materials (Bristol, England). 2016;**11**:015020

[181] Barralet J, Gbureck U, Habibovic P, Vorndran E, Gerard C, Doillon CJ. Angiogenesis in calcium phosphate scaffolds by inorganic copper ion release. Tissue Engineering. Part A. 2009;**15**:1601-1609

[182] Schamel M, Bernhardt A, Quade M, Wuerkner C, Gbureck U, Moseke C, et al. Cu^{2+}, Co^{2+} and Cr^{3+} doping of a calcium phosphate cement influences materials properties and response of human mesenchymal stromal cells. Materials Science and Engineering: C. 2017;**73**:99-110

[183] Bernhardt A, Schamel M, Gbureck U, Gelinsky M. Osteoclastic differentiation and resorption is modulated by bioactive metal ions Co^{2+}, Cu^{2+} and Cr^{3+} incorporated into calcium phosphate bone cements. PLoS One. 2017;**12**:e0182109

[184] Bergmanna C, Lindnera M, Zhangb W, Koczura K, Kirstena A, Telleb R, et al. 3D printing of bone substitute implants using calcium phosphate and bioactive glasses. Journal of the European Ceramic Society. 2010;**30**:2563-2567

[185] Trombetta R, Inzana J, Schwarz E, Kates S, Awad H. 3D printing of calcium phosphate ceramics for bone tissue engineering and drug delivery. Annals of Biomedical Engineering. 2017;**45**:23-44

[186] Ngo T, Kashani A, Imbalzano G, Nguyena K, Hui D. Additive manufacturing (3D printing): A review of materials, methods, applications and challenges. Composites Part B: Engineering. 2018;**143**:172-196

[187] Akkinenia A, Luoa Y, Schumacher M, Nies B, Lode A, Gelinsky M. 3D plotting of growth factor loaded calcium phosphate cement scaffolds. Acta Biomaterialia. 2015;**27**:264-274

[188] Luo Y, Lode A, Sonntag F, Nies B, Gelinsky M. Well-ordered biphasic calcium phosphate–alginate scaffolds fabricated by multi-channel 3D plotting under mild conditions. Journal of Materials Chemistry B. 2013;**1**:4088

Chapter 8

Amorphous Calcium Phosphate as Bioactive Filler in Polymeric Dental Composites

Diane R. Bienek, Anthony A. Giuseppetti and Drago Skrtic

Abstract

As biocompatible and osteo-inductive precursor to biological apatite formation, amorphous calcium phosphate (ACP) resorbs at the rate that closely coincides with the rate of new bone formation and is more osteo-conductive than its crystalline counterpart. In addition, in the oral environment, ACP intrinsically provides a protracted supply of the remineralizing calcium and phosphate ions needed for regeneration of mineral lost to tooth decay. These features make ACP composites a strong remineralizing tool at the site of caries attack. Our group has been on the forefront of the research on bioactive, remineralizing, polymeric ACP-based dental materials for over two decades. This entry describes methods for filler, polymer, and composite fabrication and a battery of physicochemical and biological tests involved in evaluation of ACP-based restoratives. Also presented is our most recent design of ACP remineralizing composites with added antimicrobial capability that shows promise for extended dental and, potentially, wider biomedical applications.

Keywords: amorphous calcium phosphate, bioactivity, dental composite, dental resin, remineralization

1. Introduction

Due to their abundance in nature as phosphate minerals and their existence in living organisms, calcium phosphates (CaPs) are of special significance to humans. CaPs are involved in normal (bones, teeth, antlers) as well as pathological (calcification/mineral deposition in soft tissues) mineralization. Both normal and pathological calcifications represent in vivo crystallization. In contrast, dental caries and osteoporosis are manifestations of in vivo dissolution where less soluble CaPs are being replaced by the more soluble ones. The extensive overviews of the current knowledge on CaP structures, properties, and biomedical and dental importance are provided in Refs. [1–4].

Amorphous calcium phosphates (ACPs) are unique members of CaP family with glass-like physical properties and variable chemistry [5]. Evidence of ACPs being an integral mineral component of bones and teeth is, however, ambiguous [6, 7]. Consequently, ACPs are considered the transient precursors in biomineralization [8–11] with the majority of information on their possible roles in biomineralization originating from the synthetic and/or in vitro studies [12].

The use of CaPs in dentistry is roughly a century old with the first scientific article published in 1925 [3]. Majority of the literature on dental use of CaPs focuses on crystalline CaPs. Today, two ACP-based remineralization systems have been commercialized as a toothpaste: a casein phosphopeptide-stabilized ACP and an unstabilized ACP. Other ACP applications include various biocompatible formulations where ACP acts as anticariogenic, remineralizing agent (polymeric composites, chewing gums, sugar confections, bleaching gels, and/or mouth rinses).

In this chapter, we present an overview of our group's up-to-date work on ACP-based dental composites with the emphasis on fabrication and characterization of ACP filler, fine-tuning of polymer phase, and physicochemical, mechanical, and biological evaluation of the ensuing polymeric ACP composites. Attention is also given to our most recent efforts to develop bioactive composites with both remineralizing and antimicrobial (AM) capabilities.

2. ACP-based polymeric dental composites

The approaches to synthesize ACP include precipitation from supersaturated calcium and phosphate solutions (wet synthesis), spray drying of acidified aqueous solutions of soluble CaPs, precipitation from nonaqueous solutions and solvents (sol-gel technique), and dry chemical techniques (mechanochemical methods, high-energy processing at elevated temperatures). Morphology, chemical composition, atomic structure, thermal properties, mechanical properties, and kinetics of ACP's transformation into crystalline CaPs depend on the preparation method [5]. ACPs fabricated by wet chemical method typically have a relatively constant chemical composition, suggesting the existence of well-defined structural units/clusters [12, 13]. ACPs are thermodynamically unstable in solutions and convert spontaneously into crystalline CaPs, predominantly apatite. As result of ACP to apatite transformation, both crystallinity and Ca/P ratios of the solid increase with time. The release of remineralizing ions that accompany ACP's transformation in solutions, coupled with their excellent biocompatibility and bioresorbability, makes ACP-based dental materials a powerful remineralizing tool. However, the unique benefit of ACP-based materials, i.e., their sustained remineralization capability, is compromised by the inherently low strength and toughness of ACP composites due to the uncontrolled agglomeration of ACP particles and enhanced water sorption (WS) on exposure to aqueous environment [14]. To overcome these deficiencies, we explored multiple approaches to reduce the heterogeneity of filler particle size distribution (PSD) and better control the stability of the ACP/resin interface. These attempts are described in Sections 2.1–2.3.

2.1 ACP filler: synthesis, characterization, and surface modification

ACPs utilized in our studies were synthesized according to the protocol originally proposed [15]. It was modified to include ab initio addition of cations or nonionic surfactants to Ca reactant and anionic surfactant or polymer to PO_4 reactant. In brief, ACP precipitated instantly upon mixing equal volumes of the reactant solutions {Ca reactant, 80 mmol/L $Ca(NO_3)_2$; PO_4 reactant, 54 mmol/L Na_2HPO_4 + 2 mol% $Na_4P_2O_7$ (to inhibit the precipitation of apatite simultaneously with ACP)} at 23°C and pH $\geq$ 8.5. After filtering, solid was subsequently washed with ice-cold ammoniated water, then with acetone, freeze-dried, and lyophilized. Dry, as-synthesized ACP (as-ACP) was kept in dessicator under vacuum before being subjected to additional treatments: silanization [16], grinding [17], or mechanical milling [14]. Before being utilized for composite fabrication, ACPs were characterized (validated) by multiple screenings. The amorphousness of the solids was confirmed by X-ray diffraction (XRD) analysis and Fourier-transform infrared (FTIR) spectroscopy. PSD of ACP dispersed in isopropanol was determined by laser light scattering, and their morphology/topology of gold-sputtered specimens was assessed by scanning electron microscopy (SEM). ACP's water content was measured by thermogravimetric analysis (TGA), and Ca/PO_4 ratio was calculated from atomic emission spectroscopy data obtained upon dissolving ACP powder in HCl. Typical XRD pattern, FTIR spectrum, and SEM image of ACP are presented in **Figure 1**. More detailed description of ACP preparatory protocols and filler's characterization/validation is provided [18]. The same reference is also a good source of information on methods and techniques utilized in preparation and evaluation of copolymers and composites and their physicochemical, mechanical, and biological assessments discussed in Sections 2.2 and 2.3.

Introducing additives during ACP synthesis and/or applying secondary treatments (**Table 1**) was expected to yield less clustered dry powders (more homogeneous and narrower PSD) and lead to better dispersion of ACP in the resin. Experimental details on surface modification protocols are provided in [16, 19]. More intimate ACP/resin contact is expected to enhance hydrolytic stability of ACP through lowering the WS, thus improving the mechanical performance of composites.

The coprecipitation of Ag and Fe phosphates and premature ACP to apatite conversion in the presence of Fe^{2+} and Fe^{3+} disqualified these cations from further evaluation. Results of PSD analysis and TGA results obtained with surface-modified ACPs and strength testing of model 2,2-bis[p-(2′-hydroxy-3′-methacryloxypropoxy) phenyl] propane (Bis-GMA)-based composites employing these fillers are summarized in **Figure 2**.

Regardless of the type of treatment, practically all ACPs preserved their heterogeneous PSDs. However, values of the median diameter (d_m), i.e., the midpoint of the volume size distribution, of Zn-ACP and Al-ACP were significantly lower than the d_ms of all other ACPs subjected to surface modification and marginally lower than the d_ms of ground- (g-ACP) or milled- (m-ACP) solids. By comparing the PSD histograms and the corresponding SEM images (not shown here), it is concluded that the degree of particle agglomeration in Zn-ACP and Al-ACP has been modestly reduced—the quantification of the effect was not attempted. A similar conclusion was made after examining the PSDs and SEMs of g- and m-ACP vs. the control.

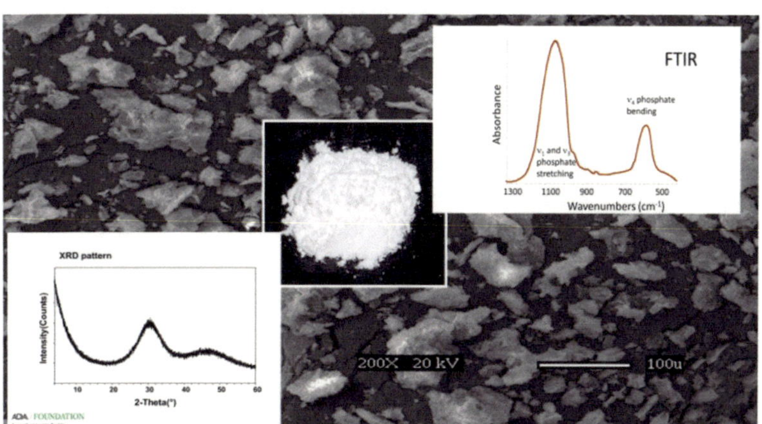

Figure 1.
Representative XRD pattern (lower left corner) and FTIR spectrum (upper right corner) of ACP powder (center) with the corresponding SEM image (background).

Additives	Expected interactions/effect(s)
Cations	
Ag^{+}, Fe^{2+}, Zn^{2+}, Al^{3+}, Fe^{3+}, Si^{4+}, Zr^{4+}	ACP/cation interactions controlled by the cation's ionic potential
Surfactants*	
Anionic-Zonyl FSP	As dispersants, surfactants prevent agglomeration of ACP particles and their coalescence with water
Nonionic-Triton100, Tween 80, Zonyl FSN	
Polymer*	
PEO (M_w 8 K, 100 K, 1000 K)	Polymer stabilizes ACP particles via multiple chelation
Treatments	
Silanization*	
(APTMS, MPTMS)	Silane agents stimulate formation of durable bonds at ACP/resin interface
Grinding	Particle size reduced by friction
Ball milling	Particle size reduced and PSD homogenized via high-energy impact and collision

Acronyms are defined in the appended list of abbreviations.

Table 1.
Surface modification of ACP: ab initio additives and secondary treatments.

The significant increase in d_m of polymer-ACPs, on the other hand, could be attributed to the effect similar to "polymer bridging" seen in apatite/high-molecular-weight (Mw) polyacrylate system [18]. In theory, poly(ethylene oxide) (PEO)/ACP interactions are controlled by the conformational changes in the adsorbed polymer, collision rate between the particles, and/or aggregate breakup due to fluid shear [20]. However, the exact controlling mechanism is yet to be determined.

The total content of surface-bound and/or structural water in all ACPs was unaffected by the treatment (on average (16.0 ± 1.2) mass%) and compared well with the control ((15.8 ± 3.9) mass%). Ideally, the lower intrinsic water content associated with less agglomerated ACP filler is desired to ensure favorable WS and ion release. WS of dental composites is generally controlled by structure/composition of the resin matrix. In case of ACP composites, the hydrophilic filler increases the amount of water absorbed. In ACP/Bis-GMA composites [21], filler/resin interface is inundated by the numerous voids that may enhance water diffusion and hydration of the filler. These processes are better controlled in m-ACP composites [14].

A simple BFS [22] screening of Bis-GMA-based ACP composites (**Figure 2**) revealed the following order of mechanical stability upon extended (up to 3 months) exposure to aqueous environment: (m-ACP ≥ g-ACP ≥ Zn-ACP = Zr-ACP = silanized ACP) ≥ control (unmodified) ACP ≥ (Si-ACP ≥ surfactant-ACPs) ≥ (Al-ACP = PEO-ACP). The observed decrease in BFS of ACP composites can, generally, be attributed to either reduction in ACP's intactness caused by spatial changes that occur in parallel to ion release, the internal ACP to apatite conversion, or increased WS. The experimental finding that, after aqueous immersion, the BFS of Al-ACP and PEO-ACP composites deteriorated independently of their PSDs (d_m of PEO-ACP was more than six times larger than that of Al-ACP, while their BFS was practically identical) suggests that fillers PSD has only a minor, if any, role in composite's ability to resist plasticization/degradation upon water exposure.

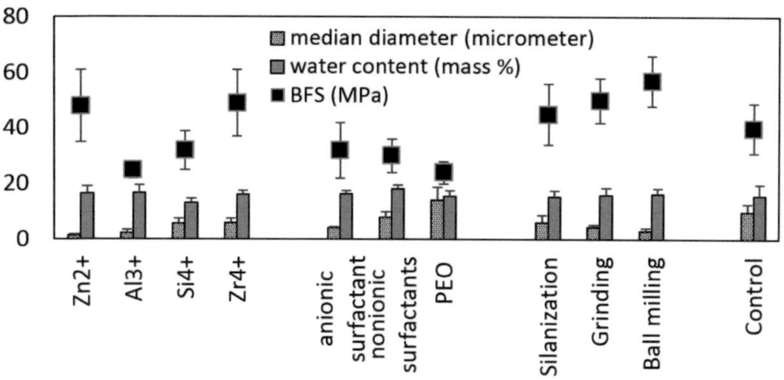

Figure 2.
Effect of additives/treatments on PSD and water content of ACP solids and the biaxial flexure strength (BFS) following the aqueous immersion of composite specimens formulated with these fillers and bis-GMA resin. Indicated are mean values of minimum three independent runs and standard deviations (SD, represented by bars). Control ACP: as-ACP made without additives or a secondary treatment.

Our findings contradict an earlier report [23] that PEO strengthened the interface with ACP interface. Better understanding of the observed behavior may require further in-depth mechanical testing and water sorption/desorption evaluations. It appears prudent to consider approaches apart from the undertaken filler treatments, such as choice of monomers utilized in resin formulations, to design composites with optimized performance.

2.2 Resin matrix: fine-tuning a means of improving copolymer properties

The experimental resins were formulated from the commercially available base, diluent, and adhesive monomers activated for light-, chemical-, and/or dual (light and chemical)-cure (LC, CC, and DC, respectively). Monomers utilized in resin formulations and the ranges of their concentrations are listed in **Table 2** (indicated acronyms will be used throughout this chapter). All commercial monomers were utilized as received, without any additional purification.

In LC formulations, all components were blended together and stirred at room temperature until reaching a uniform consistency. In CC formulations, monomers were first combined, their mixture homogenized and then divided into two equal parts by mass. The components of the CC initiating system are added separately to each monomer mixture and stirred magnetically until blended fully. In DC systems, LC and CC step were combined. LC and DC preparations required the use of yellow light to prevent the premature photopolymerization.

Copolymer specimens were prepared by packing resins into Teflon molds, covering each side of the mold with Mylar film and a glass slide, clamping assembly together, and, for LC and DC specimens, curing each side of the assembly with visible light for 2 minutes. Physicochemical tests routinely performed with copolymers involved assessment of degree of vinyl conversion (DVC; near-IR spectroscopy), WS (gravimetric measurements), and BFS.

Methacrylate-based dental resins typically entail high-viscosity base monomer (due to its larger molecular volume decreases polymerization shrinkage (PS) and enhances modulus of cured copolymer) and low-viscosity diluent monomer (due to its smaller molecular volume and greater flexibility enhances DVC and handling properties). Majority of contemporary dental resins is Bis-GMA-/TEGDMA-based. Known drawbacks of Bis-GMA-based systems are relatively low DVCs, high PS, and susceptibility to hydrolytic and enzymatic degradation upon

Component	Chemical name	Acronym	Content (mass %)
Base monomers	2,2-Bis[p-(2′-hydroxy-3′-methacryloxypropoxy) phenyl] propane	Bis-GMA	≥68.4
	Ethoxylated bisphenol A dimethacrylate	EBPADMA	≤68.4
	Urethane dimethacrylate	UDMA	≤92.4
Diluent monomers	Ethyl-α-hydroxymethyl acrylate	EHMA	≤29.2
	2-hydroxyethyl methacrylate	HEMA	≤30.4
	Poly(ethylene glycol)-extended UDMA	PEG-U	≤29.1
	Triethylene glycol dimethacrylate	TEGDMA	≤50.2
Adhesive monomers	Methacryloyloxyethyl phthalate	MEP	≤5.0
Initiators	*Light cure*		
	Camphorquinone	CQ	0.2
	Ethyl-4 N,N-dimethylamino benzoate	4EDMAB	0.8
	Chemical cure		
	Benzoyl peroxide	BPO	2
	2,2-Dihydroxyethyl-p-toluidine	DHEPT	1

Table 2.
Methacrylate monomers and the components of polymerization-initiating systems utilized to fabricate the experimental resins.

exposure to oral fluids. In our resin fine-tuning studies, we have been exploring the utility of EBPADMA and UDMA as the alternative base monomers and EHMA, HEMA, and PEG-U as the alternative diluent monomers. These studies are expected to yield valuable information on the interplay between the resin's hydrophilicity/hydrophobicity, DVC, and mechanical stability and their effects on thermodynamic stability and mechanical performance of ACP composites fabricated with these resins.

Results of DVC and BFS screenings of Bis-GMA-, EBPADMA-, and UDMA-based resins are compiled in **Figure 3**. The DVC values attained in our experimental formulations (ranging from 81.4 to 88.2%) with varying diluent and/or adhesive monomers (**Table 2**) significantly exceeded the typical DVC values reported for Bis-GMA/TEGDMA copolymers (60–70% [24]).

DVCs attained in Bis-GMA (81.4 ± 2.5%), EBPADMA (84.8 ± 5.5%), and UDMA (88.2 ± 2.2%) copolymers were significantly different ($P < 0.05$; Tukey post hoc test) only for UDMA vs. Bis-GMA systems. The BFS values, apparently increasing in going from Bis-GMA to EBPADMA to UDMA copolymers ($R^2 = 0.9685$), were significantly different ($P < 0.05$) only between UDMA and Bis-GMA copolymers. WS (data not shown) was not significantly affected by the resin's compositional changes. In all three groups, maximum WS (values ranged from 2.3 to 3.0 mass%) was achieved within 2 weeks of aqueous immersion. Slight differences between the measured WS values could be related primarily to the inclusion of more hydrophilic (HEMA, TEGDMA) or less hydrophilic (EHMA, PEG-U) monomers in the resin matrix. It is, however, highly significant that all systems maintained the WS profiles that support a sustained ion release from the composites (see Section 2.3).

2.3 ACP composites: fabrication, physicochemical, mechanical, and biological assessments

To fabricate composites, resin (60 mass%) and ACP filler (40 mass%) were combined by hand spatulation. The homogenized paste was kept overnight under vacuum to eliminate the entrapped air before being utilized for preparation of testing specimens by employing the procedures identical to those for the preparation of copolymer specimens. Besides DVC, BFS, WS, and PS testing, physicochemical and biological evaluation of composites also entailed polymerization shrinkage stress (PSS), shear bond strength (SBS) to dentin, ion release kinetics, leachability of unreacted species, remineralization efficacy, and in vitro cytotoxicity tests.

DVCs (**Figure 4**) of Zr-ACP composites formulated with various resins significantly (ANOVA, Tukey test; $P < 0.05$) increased in going from Bis-GMA to EBPADMA to UDMA matrix. Higher DVC in UDMA systems could possibly be explained by the higher reactivity of UDMA than Bis-GMA and/or EBPADMA [25]. Resin composition, however, had no effect on the BFS of composites being ~50% lower than the BFS of their copolymer counterparts (compare BFS data in **Figure 4** vs. **Figure 3**). The reduction in BFS in UDMA-based composites (1.7%) was significantly lower ($P < 0.05$) than reduction in BFS in Bis-GMA- to EBPADMA-based composites (10.0 and 4.4%, respectively). This finding suggests that random

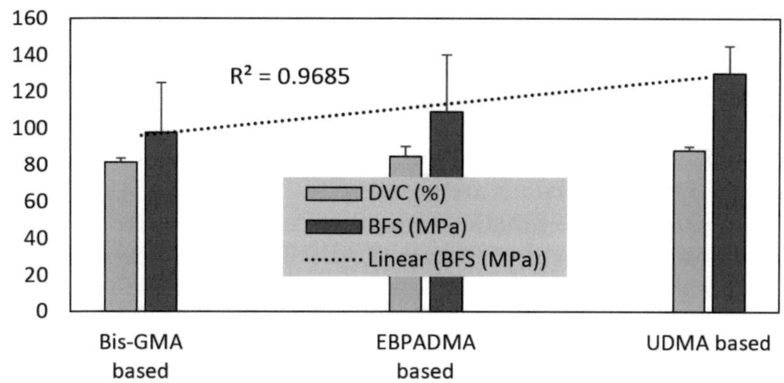

Figure 3.
DVC and BFS of LC binary, ternary, and quaternary bis-GMA-, EBPADMA-, and UDMA-based copolymers. Indicated are mean values + SD (represented by bars) for $n \geq 24$ (DVC) and $n \geq 12$ (BFS).

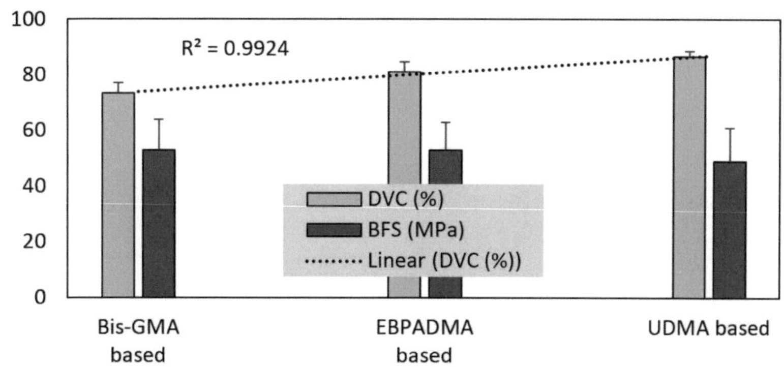

Figure 4.
DVC and BFS of LC Zr-ACP bis-GMA-, EBPADMA-, and UDMA-based composites. Indicated are mean values + SD (represented by bars) for $n \geq 24$ (DVC) and $n \geq 12$ (BFS).

distribution of Zr-ACP agglomerates [21], rather than the resin matrix composition, controls the mechanical performance of composites.

Generally, WS {(3.3–3.8) mass%}, SBS {(15.3–17.5) MPa}, and PS {(6.5–7.0) vol%} results revealed no distinguishable differences between the Bis-GMA-, EBPADMA-, and UDMA-based composites. The observed increase in WS of composites vs. their copolymer counterparts (27–43%) was due to ACP's affinity to the environmental water. It is significant that, in terms of SBS, ACP composites performed as well as Sr-glass filled composites [17], thus providing the remineralizing component to the primary restorative function without impediment of short- and midterm dentin bonding. High PS (undesirable) seen in all three experimental groups go hand in hand with the high DVCs (desirable) attained in these systems. These high PS values are likely due to the increased hydrogen bonding occurring in all experimental matrices leading to the densification of polymerization [26]. It is particularly important that, in UDMA-based composites, high PS can be offset by a significant hygroscopic expansion (HE; up to 13.6 vol%; data not shown). The compensating effect of HE on PS has been demonstrated in [27–29]. We were unable to establish any correlation between the PS and PSS in our experimental systems. Although contributions on this subject in dental literature are considerable [30–33], there is a question whether processing factors such as configuration factor (C-factor) besides the filler type and its load level, resin composition, and polymerization mode control the composite performance [34].

Comparative kinetic study of ion release from Bis-GMA- and EBPADMA-based composites showed a systematic increase in solution Ca and PO_4 concentrations with the increased filler level in the composite and with time of aqueous immersion. Values of the ion activity product (IAP) used to calculate the relative supersaturation of the solutions with respect to the stoichiometric hydroxyapatite (details on these calculations are provided in [35]) are presented in **Figure 5**. The ion release kinetics was practically identical in both types of composites. It was shown that a minimum of 35 mass% of ACP filler in composite is needed to create a sustained solution supersaturation inducive to apatite formation. The overall ion release kinetics in these systems was most likely controlled by the level of hydrophilic HEMA monomer in the resin phase. By more easily absorbing the water needed for ion diffusion, HEMA regulated the internal mineral saturation levels.

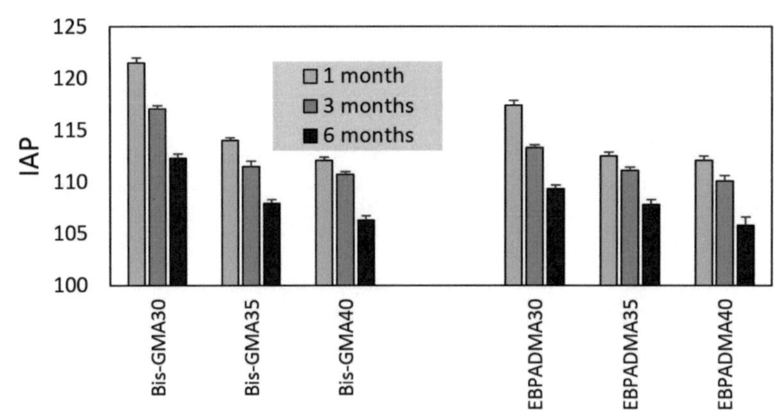

Figure 5.
Mean value + SD (indicated by bars) of the ion activity product (IAP) of the solutions containing Ca and PO_4 ions released from bis-GMA- and EBPADMA-based composites with various ACP levels (30, 35, or 40 mass%) at different times of immersion (1, 3, or 6 months) in saline solutions (23°C; continuous magnetic stirring). Number of repetitive experiments n = 3/group.

In our studies, we conveniently use the DVC as a predictor of leachability (the higher the DVC, the lower the likelihood of unreacted monomers to leach) from the restorations. To quantify leaching from the UDMA/PEG-U/HEMA/MEP (UPHM) copolymers and their ACP composites, we have employed ^{1}H-NMR spectroscopy. The accelerated leaching was mimicked by using acetone as extraction medium. Results are presented in **Figure 6**. Leaching (expressed as a mass fraction of the initial amount incorporated in the resin and normalized to the equivalent amount of the resin for composite specimens) decreased in the following order: 4EDMA B > > MEP > UDMA > PEG-U > HEMA > CQ (undetectable). Introducing ACP filler into UPHM matrix did not have an impact on the leachability profile. In highly cross-linked UPHM matrix, polymer chain mobility remained small in both copolymers and composites, thus limiting the pathways for unreacted monomers to leach out from the specimens. The disproportionate leachability of 4EDMAB, a component of the LC initiator system, is most likely due to its excess relative to CQ (CQ:4EDMAB mass ratio 1:4), which appears to be fully consumed during polymerization and is undetectable in the extracts. Since HEMA is known for its increased toxicity and adverse side effects due to metabolizing to methacrylic acid [36], it is of high significance that the level of leachable HEMA from UPHM copolymers and composites is drastically (several orders of magnitude) lower than typically seen in HEMA-containing resins and composites. In addition, levels of leachable UDMA from UPHM matrices are more than twice lower than the amounts of leached UDMA from the experimental UDMA/TEGDMA formulations [37]. In the complex, high-DVC UPHM network, the mobility of the unreacted low-molecular-weight HEMA is limited or reduced, and its elution to the environment is prevented.

In vitro cytotoxicity screening should be performed in conjunction with the leachability tests to better predict the material's impact at cellular level. To mimic early ACP composite/cell interactions, we have performed the extraction experiments and assessed cell morphology and cell viability according to the recommended standards [38–40]. The osteoblast-like MC3T3-E1cells were seeded on ethanol-sterilized material disks, incubated in the extracts for 24 h, then evaluated in situ by optical microscopy, and assessed for cytotoxicity by tetrazolium-based {3-(4,5-dimethylthiazol-2-yl)-2,5-diphenyltetrazolium bromide; MTT) assay. Detailed protocols are described in [41]. The results of the cytotoxicity evaluation of UPHM composites are presented in **Figure 7**. Cell morphology (**Figure 7**, inlets) changed from polygonal, spread (cells exposed to pure medium and extracts from UPHM copolymer and compressed ACP filler) to spherical (cells incubated in

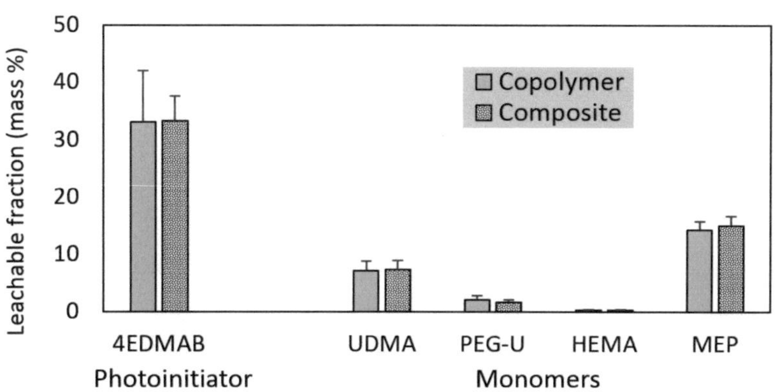

Figure 6.
Leachables (mean value + SD (represented by bars); n = 3/group) detected by ^{1}H-NMR spectroscopic method in acetone extracts from LC UPHM copolymers and their ACP composites (in composite series; values are normalized with respect to the initial resin amount).

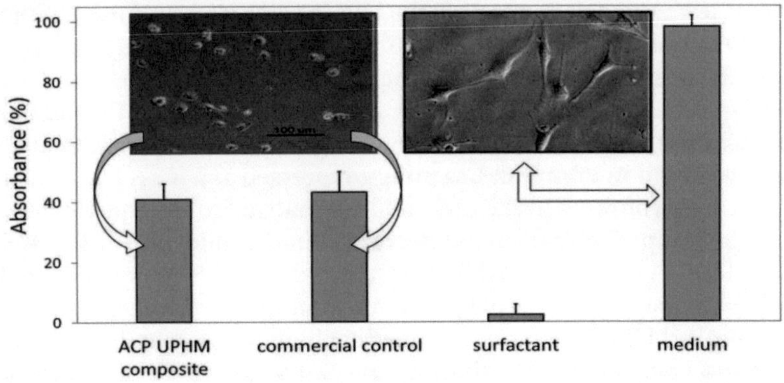

Figure 7.
Viability of osteoblast-like cells upon exposure to the extracts from ACP UPHM composite, commercial composite, medium + surfactant, and medium only. Shown are mean values = SD (indicated by bars) for three repetitive runs in each group. Optical micrographs indicating typical changes in cell morphology upon exposure to the extracts from composites (left side) and pure medium (right side).

extracts from the experimental ACP UPHM composite and commercial control). Viability of contracted cells (i.e., those exposed to extracts from both composites) was reduced approximately twofold compared to the polygonal cell incubated in medium only. Similar morphological changes and slow cell proliferation observed in bone-regenerating hydrogels [42] are reportedly linked to the alterations in cell nucleus and mineralization of osteoprogenitors [43]. While fuller understanding of the cellular effects is certainly needed, it is encouraging that reduction in cell viability seen with our experimental composites was comparable with the commercial control.

Remineralization efficacy studies were undertaken to demonstrate the ability of ACP composites to regenerate demineralized tooth structures. Details on the microradiographic in vitro evaluations of the changes in mineral content of the demineralized bovine and/or human enamel subjected to aggressive acid attacks, representative prolonged exposure in oral milieu, are provided in [44, 45], respectively. These studies indicate that lost tooth mineral is indeed regenerated upon application of ACP composites (**Figure 8**). Mineral recovery with ACP Bis-GMA-based composites significantly exceeded the remineralization effect of

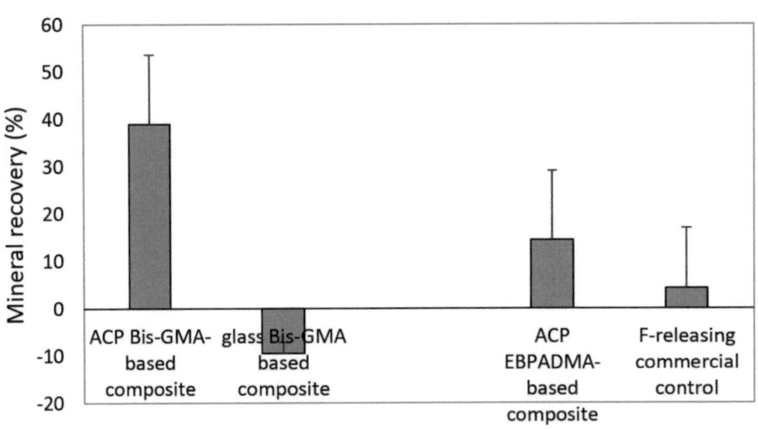

Figure 8.
Mineral recovery (or loss) following the pH regimens representing pH cycling in oral environment. Shown are mean values ± SD (indicated by bars) obtained from a minimum of eight micrographic images/experimental group.

the glass-filled control. On average, more than threefold increase of mineral content attained with ACP EBPADMA-based composites compared to F-releasing control was, however, statistically insignificant due to a large data scattering. It is particularly important that the regenerative action took place throughout the lesions in contrast to F-supported repair which is typically limited to a subsurface region.

2.4 Adding antimicrobial functionality to dental restoratives

To prolong service life, the next generation of polymeric dental restoratives is envisioned to possess bioactive properties. Today, developments of dental materials with embedded AM properties are relatively frequently reported. One of the first quaternary ammonium (QA) methacrylates integrated into Bis-GMA/TEGDMA matrix has been reported 20 years ago [46]. More recently, two polymerizable ionic dimethacrylates (2-(methacryloyloxy)-N-(2-(methacryloyloxy)ethyl)-N,N-dimethylethan-1-aminium bromide (IDMA1) and N,N′-([1,1′-biphenyl]-2,2′-diylbis(methylene))bis(2-(methacryloyloxy)-N,N-dimethylethan-1-aminium) bromide (IDMA2)) were proposed for use in dental applications [47, 48]. Purity of the synthesized materials was not reported, and the structural characterizations appeared incomplete. The IDMAs have been validated by nuclear magnetic resonance, mass spectroscopy, and FTIR spectroscopies [49, 50]. These studies also included direct contact cytotoxicity testing at biologically relevant concentrations. To date, AM assessments of the fully characterized IDMAs are remaining.

For the advancement of Class V restoratives, our goal was to evaluate the AM properties of purified/validated IDMA1 and IDMA2 integrated in dental copolymers, using *Streptococcus mutans* (planktonic and biofilm).

IDMA copolymer disks (6 mm diameter) were fabricated by adding IDMA1 or IDMA2 (10 wt %) to light-activated UDMA/PEG-U/EHMA (hereafter UPE) resin. For both planktonic and biofilm testing, UPE disks were used as a negative control group. After fabrication, all copolymer disks were subjected to aqueous extraction (72 h, 37°C). After drying, specimens were sterilized using an Anprolene gas sterilization chamber (Andersen Products, Inc., Haw River, NC). Prior to bacterial testing, specimens were degassed for ≥5 days.

A bioluminescent *S. mutans* strain JM10 [51] (derivative of wild-type UA159) was used to assess the antibacterial properties, the IDMAs, employing a real-time bioluminescent assay developed by Florez et al. [52]. In UPE resin matrix, 10% IDMA1 reduced ($P \leq 0.005$) the colonization of *S. mutans* biofilms threefold (**Figure 9**). The initial report describing the synthesis of IDMA1 and incorporation into Bis-GMA-TEGDMA resin conducted *S. mutans* testing with phosphate-buffered saline (i.e., no extrinsic proteins) [48]. Therein, they report reduced bacterial colonization for IDMA1 concentrations of 10, 20, or 30%. In the presence [53, 54] or absence [55] of ACP, others have demonstrated that IDMA1 fabricated with Bis-GMA/TEGDMA results in a modest (1.4–1.7-fold decrease in metabolic activity of biofilm) AM function in a protein-rich environment. To enhance AM activity, IDMA1 in combination with silver nanoparticles and/or nano-ACP was integrated into dental materials [53–56].

AM activity of IDMA2 has not been previously reported. Compared to the UPE control resin, our experiments demonstrated that IDMA2-UPE copolymer reduced ($P \leq 0.005$) the colonization of *S. mutans* biofilms 1.6-fold (**Figure 9**). However, the AM activity of IDMA2 was less ($P \leq 0.05$) than that observed with IDMA1. This difference may be due to steric hindrance resulting from the presence of aromatic rings in IDMA2. Further, it is noteworthy that IDMA2 can exist as a mixture of conformers. Thus, it is conceivable that the conformer (with decreased steric

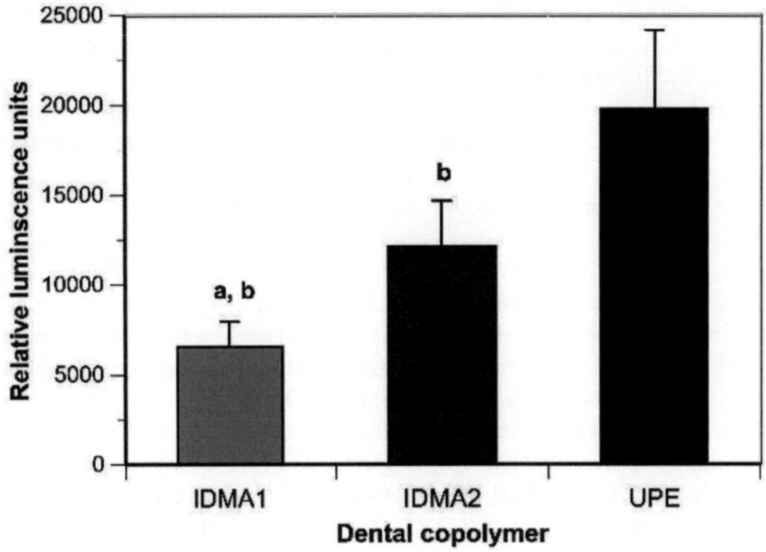

Figure 9.
Streptococcus mutans biofilm growth by the experimental IDMAs-UPE (10 wt %) copolymers compared to UPE resin. Presented are mean values ± SD (indicated by bars) of 5 specimens/experimental group. $^{a}P \leq 0.05$ *compared to IDMA2;* $^{b}P \leq 0.005$ *compared to UPE resin.*

hindrance) could exhibit greater AM activity. Future research that enables preparations, enriched for each conformer, is warranted to determine whether AM activity of IDMA2 can be enhanced.

For planktonic bacterial testing, *S. mutans* UA-159 (ATCC® 700610) cultures were established in Todd Hewitt broth (THB). Copolymer disks were seeded with *S. mutans* suspension (~3×10^7 colony-forming units/disk). To maximize the contact between the copolymer and bacteria, a second disk was placed atop. This assembly was incubated 2 h at 37°C in a 5% CO_2 environment. The samples were then placed in 1 ml of THB and utilized to prepare 10-fold serial dilutions. Of the resulting suspensions, 100 μl were distributed onto the surface of THB agar plates. Colony-forming units were the unit of measure used to approximate the number of viable *S. mutans* cells after exposure to the IDMAs-UPE copolymers. Among the IDMA groups, the number of colony-forming units observed was not statistically different from one another or the UPE control resin (data not shown).

Altogether, the full AM potential of the IDMAs has not yet been realized. For example, the current IDMA restorative material formulations are likely to have charges randomly distributed throughout the material. It is conceivable that chemical engineering advances (i.e., material gradient or layers) could favor the orientation of the N^+ charges while maintaining the material mechanical properties. Secondly, previous reports demonstrated that the presence of proteins dampens the AM capability of QA methacrylates (reviewed by [57]). To overcome this problem, a commercially available protein repellent, 2-methacryloyloxyethyl phosphorylcholine (MEPC), has been integrated into AM dental material [58]. However, because of possible binding of the remineralizing calcium ions by MEPC [59], MEPC may have a deleterious effect on the remineralization capacity of the restorative material. Generally, to date, there is a paucity of information regarding the nature of proteins binding to the surface of QA methacrylates. Elucidation of the protein-material interactions would yield valuable information to develop a strategy to maximize AM efficacy of materials with a charge-based AM mechanism of action. Further, these data could facilitate the advancement of materials, modified to target specific proteins (i.e., via molecular imprinting).

3. Conclusions

Bioactive, ACP-based polymeric dental composites ACP have tremendous appeal due to their intrinsic ability to regenerate lost tooth mineral. ACP composites release remineralizing Ca and PO_4 ions in a sustained fashion, thus providing a long-term protection against acid challenges in the oral environment. However, with respect to the mechanical strength and toughness, ACP-based composites are inferior to silanized glass-filled counterparts and can only be used in nonstress-bearing applications. The uncontrolled ACP particle agglomeration typically leads to the poor ACP filler/resin interactions that destabilize filler/polymer interface. A desired, more homogeneous PSD of ACP filler is achievable via mechanical treatments, preferably high-energy milling. Surface modifications of ACP were proven less effective in that respect. With HEMA as diluent monomer, LC Bis-GMA-, EBPADMA-, and UDMA-based copolymers and their ACP composites typically attain high-DVC values that would suggest a minimal leaching of unreacted monomeric species from these materials. Expectedly, these high DVCs are accompanied with high PS. In UDMA-based systems, high PS is likely to be offset by a significant HE. High remineralization capabilities are achieved with EBPADMA or UDMA as base monomers and HEMA or EHMA plus PEG-U as diluent monomers. The most recent efforts in our group focus on design of bifunctional, remineralizing, and AM ACP composites that may find utility as new Class V restoratives. These materials are currently being assessed for their in vitro cytotoxicity and AM properties. The evaluations coupled with the leachability studies need to be completed before embarking on animal testing and/or human clinical trials involving AM ACP dental composites.

Acknowledgements

The work presented in this chapter was supported by the National Institute of Dental and Craniofacial Research (grants DE 13169 and DE 26122), American Dental Association (ADA), and ADA Foundation. Special thanks to collaborators (Sharukh S. Khajotia, Fernando L. Esteban Florez, and Rochelle D. Hiers) at the University of Oklahoma Health Sciences Center, College of Dentistry, for conducting biofilm testing. Donation of the monomers by Esstech, Essington, PA, is gratefully acknowledged.

Conflict of interest

Authors declare no conflict of interest.

A. Appendix: list of abbreviations

ACP	amorphous calcium phosphate
ADA	American Dental Association
ADAF	American Dental Association Foundation
Al-ACP	aluminum-modified ACP
AM	antimicrobial
APTMS	3-aminopropyltrimethoxysilane
as-ACP	as synthesized ACP
BFS	biaxial flexure strength

Bis-GMA	2,2-bis[p-(2′-hydroxy-3′-methacryloxypropoxy) phenyl] propane
BPO	benzoyl peroxide
CaP	calcium phosphate
CC	chemical-cure
CQ	camphorquinone
DC	dual-cure
DHEPT	2,2-dihydroxyethyl-p-toluidine
d_m	median diameter
DVC	degree of vinyl conversion
EBPADMA	ethoxylated bisphenol A dimethacrylate
4EDMAB	ethyl-4 N,N-dimethylamino benzoate
EHMA	ethyl-α-hydroxymethyl acrylate
FTIR	Fourier-transform infrared
g-ACP	ground ACP
HE	hygroscopic expansion
HEMA	2-hydroxyethyl methacrylate
IAP	ion activity product
IDMA	ionic dimethacrylate
IDMA1	2-(methacryloyloxy)-N-(2-(methacryloyloxy)ethyl)-N,N-dimethylethan-1-aminium bromide
IDMA2	N,N′-([1,1′-biphenyl]-2,2′-diylbis(methylene))bis(2-(methacryloyloxy)-N,N-dimethylethan-1-aminium) bromide
LC	light-cure
m-ACP	milled ACP
MEP	methacryloyloxyethyl phthalate
MEPC	2-methacryloyloxyethyl phosphorylcholine
MPTMS	methacryloxypropyltrimethoxy silane
MTT	3-(4,5-dimethylthiazol-2-yl)-2,5-diphenyltetrazolium bromide assay
Mw	molecular weight
n	number of specimens/experimental runs
PEG	poly(ethylene glycol)
PEG-U	PEG-extended UDMA
PEO	poly(ethylene oxide)
PS	polymerization shrinkage
PSD	particle size distribution
PSS	polymerization shrinkage stress
QA	quaternary ammonium
SBS	shear bond strength
SEM	scanning electron microscopy
SD	standard deviation
Si-ACP	silica-modified ACP
TEGDMA	triethylene glycol dimethacrylate
TGA	thermogravimetric analysis
THB	Todd Hewitt broth
Triton 100	$C_{14}H_{22}O(C_2H_4O)_{10}H$; nonionic surfactant
Tween 80	$C_{24}H_{43}O_6(C_2H_4O)_{20}H$; nonionic surfactant
UDMA	urethane dimethacrylate
UPE	UDMA/PEG-U/EHMA resin
UPHM	UDMA/PEG-U/HEMA/MEP resin
XRD	X-ray diffraction
WS	water sorption
Zn-ACP	zinc-modified ACP

Zr-ACP	zirconia-modified ACP
Zonyl FSN	$F(C_2F_4)_x(C_2H_4O)_{1+y}H$; nonionic surfactant (x = 1–9; y = 0–25)
Zonyl FSP	$F(C_2F_4)_x(C_2H_4O)_yHP(O)(ONH_4)_z$; anionic surfactant (x = 1–7; (y + z) = 3)

Author details

Diane R. Bienek, Anthony A. Giuseppetti and Drago Skrtic*
Research Division, American Dental Association Foundation, Frederick, MD, USA

*Address all correspondence to: drago.skrtic@nist.gov

References

[1] Dorozhkin SV. Calcium orthophosphates ($CaPO_4$): Occurrence and properties. Progress in Biomaterials. 2016;**5**:9-70. DOI: 10.1007/s40204-015-0045-z

[2] Eliaz N, Metoki N. Calcium phosphate bioceramics: A review of their history, structure, properties, coating technologies and biomedical applications. Materials. 2017;**10**:334. DOI: 10.3390/ma10040334

[3] Dorozhkin SV. Calcium orthophosphates (Ca_3PO_4) and dentistry. Bioceramics Development and Applications. 2016;**6**(2). DOI: 10.4172/2090-5025: 1000096

[4] Bienek DR, Skrtic D. Utility of amorphous calcium phosphate-based scaffolds in dental/biomedical research. Biointerface Research in Applied Chemistry. 2017;**7**(1):1989-1994

[5] Dorozhkin SV. Amorphous calcium orthophosphates: Nature, chemistry and biomedical applications. International Journal of Materials and Chemistry. 2012;**2**(1):19-46. DOI: 10.5923/j.ijmc20120201.04

[6] Glimcher MJ, Bonar LC, Grynpas MD, Landis WJ, Roufosse AH. Recent studies of bone mineral: Is the amorphous calcium phosphate theory valid? Journal of Crystal Growth. 1981;**53**:100-119. DOI: 10.1016/9922-0248(81)90058-0

[7] Grynpas MD, Bonar LC, Glimcher MJ. Failure to detect an amorphous calcium phosphate solid phase in in bone mineral: A radial distribution function study. Calcified Tissue International. 1984;**36**:291-301. DOI: 10.1007/BF0240533

[8] Weiner S, Sagi I, Addadi L. Choosing the crystallization path less travelled. Science. 2005;**309**:1027-1028. DOI: 10.1126.science.1114920

[9] Weiner S. Transient precursor strategy in mineral formation of bone. Bone. 2006;**39**:431-433. DOI: 10.1016/j.bone2006.02.058

[10] Mahamid J, Sharir A, Addadi L, Weiner S. Amorphous calcium phosphate is a major component in forming fin bones of zebrafish: Indications for an amorphous precursor phase. Proceedings of the National Academy of Sciences of the United States of America. 2008;**105**:12748-12753. DOI: 10.1073/pnas.0806869105

[11] Beniash A, Metzler RA, Lam RSK, Gilbert PUPA. Transient amorphous calcium phosphate in forming enamel. Journal of Structural Biology. 2009;**166**:133-143. DOI: 10.1016/j.jsb2009.02.001

[12] Eanes ED. Amorphous calcium phosphate. In: Chow LC, Eanes ED, editors. Monographs in Oral Science. Vol. 18. Basel: Karger; 2001. pp. 130-147. ISBN: 9783318007046 3318007048

[13] Heughebaert JC, Montel G. Conversion of amorphous calcium phosphate into apatitic tricalcium phosphate. Calcified Tissue International. 1982;**34**:S103-S108. DOI: 10.1007/BF02426637

[14] O'Donnell JNR, Antonucci JM, Skrtic D. Illuminating the role of agglomerates on critical physicochemical properties of amorphous calcium phosphate composites. Journal of Composite Materials. 2008;**42**:2231-2246. DOI: 10.1177/0021998308094797

[15] Eanes ED, Gillessen IH, Posner AS. Intermediate states in the precipitation of hydroxyapatite. Nature.

1965;**208**(5008):365-367. DOI: 10.1038/208365a0

[16] Antonucci JM, Skrtic D. Bioactive and biocompatible polymeric composites based on amorphous calcium phosphate. In: Ramalingam M, Tiwari A, Ramakrishna S, Kobayashi H, editors. Integrated Biomaterials for Medical Applications. Salem: Scrivener Publishing; 2012. pp. 67-119. DOI: 10.1002/9781118482513.ch3

[17] O'Donnell JNR, Skrtic D. Degree of vinyl conversion, polymerization shrinkage and stress development in experimental endodontic composite. Journal of Biomimetics Biomaterials and Tissue Engineering. 2009;**4**:1-12. DOI: 10.4028.www.scientific.net/JBBTE.4.1

[18] Skrtic D, Antonucci JM. Dental composites: Bioactive polymeric amorphous calcium phosphate-based. In: Encyclopedia of Biomedical Polymers and Polymeric Biomaterials. New York: Taylor and Francis; 2015. pp. 2443-2262. DOI: 10.1081/E-EBPP-120051063

[19] Ofir PBY, Govrin-Lipman R, Garti N, Furedi-Milhofer H. The influence of polyelectrolytes on the formation and transformation of amorphous calcium phosphate. Crystal Growth & Design. 2004;**4**(1):177-183. DOI: 10.1021/cg034148g

[20] Antonucci JM, Liu DW, Skrtic D. Amorphous calcium phosphate-based composites: Effect of surfactants and poly (ethylene oxide) on filler and composite properties. Journal of Dispersion Science and Technology. 2007;**28**(5):819-824. DOI: 10.1080/01932690701346255

[21] Skrtic D, Antonucci JM, Eanes ED, Eidelman N. Dental composites based on hybrid and surface-modified amorphous calcium phosphates. Biomaterials. 2004;**25**:1141-1150. DOI: 10.1016/j.biomaterials.2003.08.001

[22] ASTMF394-78 (re-approved 1991). Standard test method for biaxial flexure strength of ceramic substrates

[23] Li Y, Weng W, Cheng K, Du P, Shen G, Wang J, et al. Preparation of amorphous calcium phosphate in the presence of poly (ethylene glycol). Journal of Materials Science Letters. 2003;**22**(14):1015-1016. DOI: 10.1023/A:1024741426069

[24] Marovic D, Panduric V, Tarle Z, Ristic M, Sariri K, Demoli N, et al. Degree of conversion and microhardness of dental composite resin materials. Journal of Molecular Structure. 2013;**1044**:299-302. DOI: 10.1016/j.molstruc.2012.10.062

[25] Morgan DR, Kalachandra S, Shobha HK, Gunduz N, Stejskal EO. Analysis of a dimethacrylate copolymer (bis-GMA and TEGDMA) network by DSC and ^{13}C solution and solid-state NMR spectroscopy. Biomaterials. 2000;**21**(18):1897-1903. DOI: 10.1016/S142-9612(00)00067-3

[26] Skrtic D, Antonucci JM. Dental composites based on amorphous calcium phosphate–resin composition/physicochemical properties study. Journal of Biomaterials Applications. 2007;**21**(4):375-393. DOI: 10.1177/0885328206064823

[27] Momoi Y, Mccabe JF. Hygroscopic expansion of resin based composites during 6 months water storage. British Dental Journal. 1994;**176**(3):91-96. DOI: 10.1038/sj.bdj.4808379

[28] Huang C, Tay FR, Cheung GSP, Kei LH, Wei SHY, Pashley DH. Hygroscopic expansion of compomer and a composite on artificial gap reduction. Journal of Dentistry. 2002;**30**(1):11-19. DOI: 10.1016/S0300-5712(01)00053-7

[29] Suiter EA, Watson LE, Tantbirojn C, Lou JSB, Versluis A. Effective expansion: Balance between shrinkage

and hygroscopic expansion. Journal of Dental Research. 2016;**95**(5):543-549. DOI: 10.1177/0022034516633450

[30] Braga LL, Ferracane JL. Contraction stress related to degree of conversion and reaction kinetics. Journal of Dental Research. 2002;**81**(2):114-118. DOI: 10.1177/0810114

[31] Ferracane JL. Developing a more complex understanding of stresses produced in dental composites during polymerization. Dental Materials. 2005;**21**(1):36-42. DOI: 10.1016/j.dental.2004.10.004

[32] Kleverlaan CJ, Feizler AJ. Polymerization shrinkage and contraction stress of dental resin composites. Dental Materials. 2005;**21**(1):1150-1157. DOI: 10.1016/j.dental.2005.02.004

[33] Choi KK, Condon JR, Ferracane JL. The effects of adhesive thickness on polymerization contraction stress of composite. Journal of Dental Research. 2000;**79**(3):812-817. DOI: 10.1177/00220345000790030501

[34] Antonucci JM, Giuseppetti AA, O'Donnell JNR, Schumacher GE, Skrtic D. Polymerization stress development in dental composites: Effect of cavity design factor. Materials. 2009;**2**(1):169-180. DOI: 10.3390/ma2010169

[35] O' Donnell JNR, Langhorst SE, Fow MD, Skrtic D. Light-cured dimethacrylate-based resins and their composites; Comparative study of mechanical strength, water sorption and ion release. Journal of Bioactive and Compatible Polymers. 2008;**23**:207-226. DOI: 10.1177/0883911508089932

[36] Durner J, Walther UI, Zaspel J, Hickel R, Reich FX. Metabolism of TEGDMA and HEMA in human cells. Biomaterials. 2010;**31**:818-823. DOI: 10.1016/j.biomaterials.2009.09.097

[37] Floyd CJE, Dickens SH. Network structure of bis-GMA- and UDMA-based resin systems. Dental Materials. 2006;**22**:1143-1149. DOI: 10.1016/j.dental.2005.10.009

[38] International Standard ISO10993-5. Biological evaluation of medical devices- Part 5. Tests for cytotoxicity; In vitro methods. Geneva, Switzerland: International Organization for Standardization; 1992

[39] International Standard ISO7405. Dentistry–preclinical evaluation of biocompatibility of medical devices used in dentistry. Test methods for dental materials. Geneva, Switzerland: International Organization for Standardization; 1997

[40] ANSI/ADA Specification No. 41. Recommended standard practices for biological evaluation of dental materials. 2005

[41] Simon CG, Antonucci JM, Liu DW, Skrtic D. *In vitro* cytotoxicity of amorphous calcium phosphate composites. Journal of Bioactive and Compatible Polymers. 2005;**20**(3):279-295. DOI: 10.1177/0883911505061854

[42] Chatterjee K, Lin-Gibson S, Wallace WE, Parekh SH, Lee YJ, Cicerone MT, et al. The effect of 3D hydrogel scaffold modulus on osteoblast differentiation and mineralization revealed by combinatorial screening. Biomaterials. 2010;**39**(19):5051-5062. DOI: 10.1016/j.biomaterials.2010.03.024

[43] Dalby MJ, Biggs MJP, Gadegaard N, Kalna G, Wilkinson CDW, Oreffo ROC. The control of human mesenchymal cell differentiation using nanoscale symmetry and disorder. Nature Materials. 2007;**6**(12):997-1003. DOI: 10.1038/nmat2013

[44] Skrtic D, Hailer AW, Takagi S, Antonucci JM, Eanes ED. Quantitative assessment of the efficacy of amorphous

calcium phosphate/methacrylate composites in remineralizing caries-like lesions artificially produced in bovine enamel. Journal of Dental Research. 1996;**75**(9):1679-1686. DOI: 10.1177/00220345596075091001

[45] Langhorst SE, O'Donnell JNR, Skrtic D. *In vitro* remineralization effectiveness of polymeric ACP composites: Quantitative micro-radiographic study. Dental Materials. 2009;**25**:884-891. DOI: 10.1016/j.dental.2009.01.094

[46] Imazato S, Tarumi H, Kato S, Ebisu S. Water sorption and color stability of composites containing the antibacterial monomer MDPB. Journal of Dentistry. 1999;**27**(4):279-283

[47] Antonucci JM. Polymerizable biomedical composition. US patent 8217081; July 2012

[48] Antonucci JM, Zeiger DN, Tang K, Lin-Gibson S, Fowler BO, Lin NJ. Synthesis and characterization of dimethacrylates containing quaternary ammonium functionalities for dental applications. Dental Materials. 2012;**28**(2):219-228. DOI: 10.1016/j.dental.2011.10.004

[49] Bienek DR, Frukhtbeyn SA, Giuseppetti AA, Okeke UC, Pires RM, Antonucci JM, et al. Ionic dimethacrylates for antimicrobial and remineralizing dental composites. Annals of Dentistry and Oral Disorders. 2018:108. PMID:30854515

[50] Bienek DR, Frukhtbeyn SA, Giuseppetti AA, Okeke UC, Skrtic D. Antimicrobial monomers for polymeric dental restoratives: Cytotoxicity and physicochemical properties. Journal of Functional Biomaterials. 2018;**9**(1):20. DOI: 10.3390/jfb9010020

[51] Merritt J, Kreth J, Qi F, Sullivan R, Shi W. Non-disruptive, real-time analyses of the metabolic status and viability of *Streptococcus mutans* cells in response to antimicrobial treatments. Journal of Microbiological Methods. 2005;**61**(2):161-170. DOI: 10.1016/j.mimet.2004.11.012

[52] Esteban Florez FL, Hiers RD, Smart K, Kreth J, Qi F, Merritt J, et al. Real-time assessment of *Streptococcus mutans* biofilm metabolism on resin composite. Dental Materials. 2016;**32**(10):1263-1269. DOI: 10.1016/j.dental.2016.07.010

[53] Cheng L, Weir MD, Xu HH, Antonucci JM, Kraigsley AM, Lin NJ, et al. Antibacterial amorphous calcium phosphate nanocomposites with a quaternary ammonium dimethacrylate and silver nanoparticles. Dental Materials. 2012;**28**(5):561-572. DOI: 10.1016/j.dental.2012.01.005

[54] Cheng L, Zhang K, Zhou CC, Weir MD, Zhou XD, Xu HH. One-year water-ageing of calcium phosphate composite containing nano-silver and quaternary ammonium to inhibit biofilms. International Journal of Oral Science. 2016;**8**(3):172-181. DOI: 10.1038/ijos.2016.13

[55] Li F, Weir MD, Chen J, Xu HH. Comparison of quaternary ammonium-containing with nano-silver-containing adhesive in antibacterial properties and cytotoxicity. Dental Materials. 2013;**29**(4):450-461. DOI: 10/1016/j.dental.2013.01.012

[56] Melo MA, Guedes SF, Xu HH, Rodrigues LK. Nanotechnology-based restorative materials for dental caries management. Trends in Biotechnology. 2013;**31**(8):459-467. DOI: 10.1016/j.tibitech.2013.05.010

[57] Imazato S. Bio-active restorative materials with antibacterial effects; new dimension of innovation in restorative dentistry. Dental Materials Journal. 2009;**28**(1):11-19. DOI: 10.4012/dmj.28.11

[58] Wang JJ, Liu F. Photoinduced graft polymerization of 2-methacryloyloxyethyl phosphorylcholine on silicone hydrogels for reducing protein adsorption. Journal of Materials Science. Materials in Medicine. 2011;**22**(12):2651-2657. DOI: 10.1007/s10856-011-4452-y

[59] Zhang N, Zhang K, Xie X, Dai Z, Zhao Z, Imazato S, et al. Nanostructured polymeric materials with protein-repellent and anti-caries properties for dental applications. Nanomaterials. 2018;**8**:393. DOI: 10.3390/nano8060393

Chapter 9

Structural and Calorimetric Studies of Zinc, Magnesium and Manganese Based Phosphate and Phosphate-Silicate Glasses

Refka Oueslati Omrani, Mohamed Jemal, Ismail Khattech and Ahmed Hichem Hamzaoui

Abstract

Glasses of the (50-x/2)Na_2O-xMO-(50-x/2)P_2O_5 (M = Zn, Mg or Mn) ($0 \leq x \leq 33$ mol%), (50-x)Na_2O-xMO-50P_2O_5 (M = Zn, Mn) ($0 \leq x \leq 33$ mol%), and (0.9-x) $NaPO_3$-xSiO_2-0.1ZnO ($0 \leq x \leq 0.1$ mol) were prepared by the melt quenching technique. Samples were investigated by means of X-ray diffraction, Archimede's method, ellipsometry, Fourier-transformed infrared (FTIR), Raman, 31P solid state magic angle spinning nuclear magnetic resonance (MAS-NMR), UV-visible spectroscopy and calorimetry. For zinc, manganese and magnesium phosphate glasses, the increase in density with the addition of MO oxide suggests the compactness of the vitreous network. For zinc phosphate silicate glasses, the variations of density and refractive index were attributed to the structural changes when SiO_2 oxide is progressively introduced. The increase in the glass transition temperature (Tg) reflects an increase in the cross-link strength of the structure as MO and SiO_2 oxides are gradually incorporated. For all glass composition, spectroscopic investigations revealed the depolymerization of metaphosphate chains (Q^2) allowing the formation of phosphate dimers (Q^1). Calorimetric dissolution shows that the dissolution is endothermic for lower MO content and become exothermic when x rises. For (50-x/2)Na_2O-xZnO-(50-x/2)P_2O_5 ($0 \leq x \leq 33$ mol%) glasses, the formation enthalpy increases with the incorporation of ZnO oxide.

Keywords: phosphate glasses, phosphate-silicate glasses, depolymerization, modifier oxide, calorimetric dissolution, formation enthalpy, thermochemical cycle

1. Introduction

Over the past several decades, great interests have been considered for phosphate glasses due to their superior physical properties which impart them a several advantages over conventional silicate and borate glasses.

Phosphate-based glasses are an interesting class in the world of glass and glass ceramics owing to their higher thermal expansion, lower melting and softening temperature, higher ultra-violet transmission and optical characteristic. Phosphate glasses have potential of applications in medicine, biology, batteries, laser technology, electronic, telecommunication [1–44].

In recent times, phosphate glasses have received considerable interest as a result of the synthesis of new glass composition with high chemical stability. The improvement of this quality induces the application of phosphate glasses in numerous fields of materials science, such as fast ionic conductors, semiconductors, photonic materials, hermetic seals, rare-earth in host solid state lasers and biomedical materials [1–44].

Thus, phosphate and silicate glasses are the most important materials which can extensively be used for laser sources and fiber amplifiers [3].

The basic structure unit of the phosphate network is based on corner-sharing PO_4 units which form chains and rings or isolated groups PO_4 [6].

The covalence of the bridging oxygen atoms is responsible for the formation of phosphate groups [18, 27].

The structure of the three dimensional network is obtained by linking three oxygen atoms with others PO_4 tetrahedrons. The metaphosphate groups contain two covalent bridging oxygen atoms. Whereas, the pyrophosphate groups are formed by bending only single oxygen atom with other tetrahedral site.

Recently, the structure of phosphate glass is describes using the O/P ratio [13, 18, 19, 27]. Furthermore, many investigators used the O/P ratio in order to classify the distribution of phosphate groups in the vitreous network. **Table 1** reports the classification of phosphate glasses as a function of O/P ratio [45].

For $2.5 < O/P < 3$, the glass network is described by the distribution of ultraphosphate groups [19].

The metaphosphate groups are obtained for O/P ratio equal to 3. The glass network is described by the connection of PO_4 tetrahedral anions with neighbors in order to form chains and rings.

For polyphosphate glasses, the O/P ratio is between 3 and 3.5. The structure is described by chains formed by PO_4 tetrahedral anions joined with others.

For O/P = 3.5, the structure is obtained by forming the phosphate dimers such as pyrophosphate groups in which two PO_4 tetrahedral shared one bridging oxygen [19]. For $3.5 < O/P < 4$, the isolated PO_4^{3-} units are formed such as orthophosphate.

The increase of O/P ratio induces the depolymerization of phosphate groups which suggests the shortening of the average chain length [19].

The network connectivity of phosphate compound is conventionally expressed as Q^n tetrahedral sites (n = 0…3), when n is the number of bridging oxygen (BO) per Q unit to neighbor phosphate tetrahedral [2, 23, 28]. Q^0 represents orthophosphates (PO_4^{3-}), Q^3 is pure P_2O_5 and Q^2 (metaphosphates) and Q^1 (pyrophosphates)

O/P	Classification	Q^n
2.5-3	Ultraphosphates	Q^3
3	Metaphosphates	Q^2
>3	Polyphosphates	$Q^2 + Q^1$
3.5	Pyrophosphates	Q^2
>3.5	Orthophosphates	Q^0
O: oxygen atom	**P**: Phosphorus atom	**Q**: Phosphorus tetrahedral sites

Table 1.
Classification of phosphate glasses as a function of O/P ratio and Qn tetrahedral sites.

are intermediate structures [11, 13, 16, 19, 20]. **Figure 1a** shows the nomenclature of phosphate groups as Q^n tetrahedral sites also with the variation of an O/P ratio [45].

For silicon-oxygen networks, n varies between 0 and 4, where Q^0 represents orthosilicates (SiO_4^{4-}), Q^4 is pure SiO_2 and Q^3, Q^2 and Q^1 represents the intermediate silicate structures [2, 20].

Glasses results from many possible combinations of network-forming oxides together with one or several modifier or intermediate oxides which lead to a special physical properties [22].

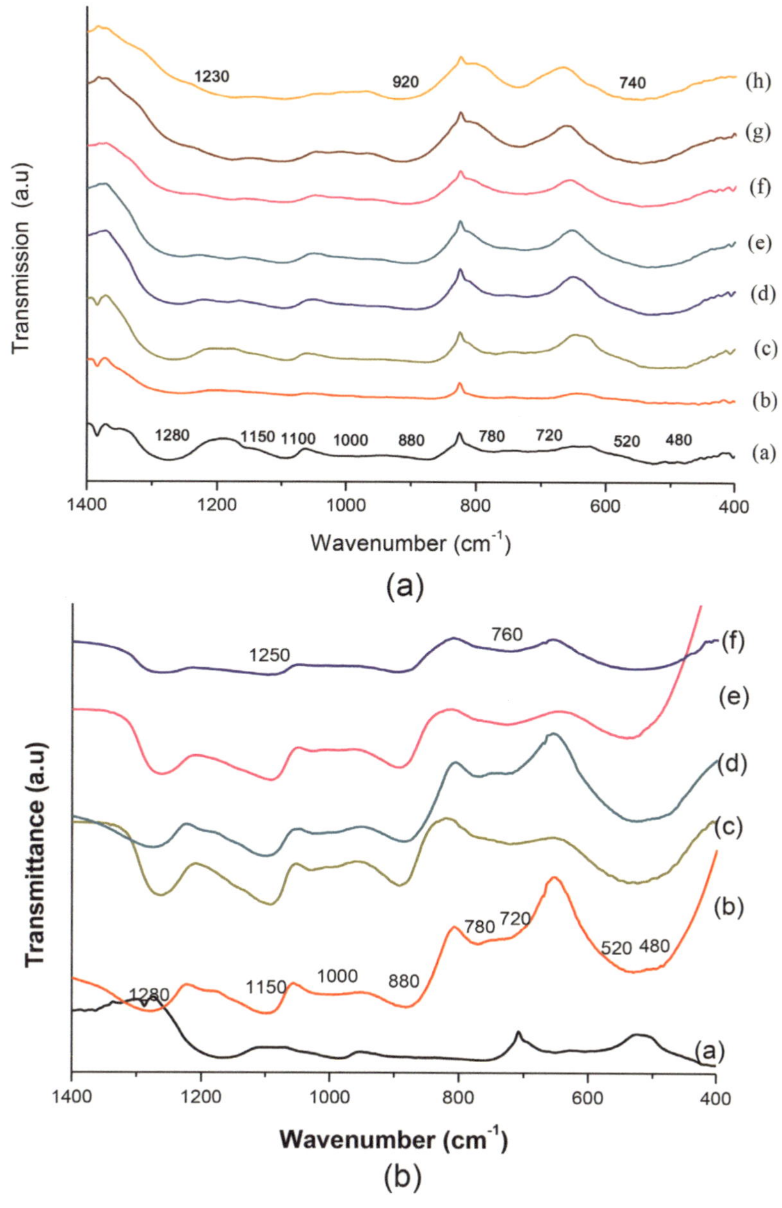

Figure 1.
a. Infrared spectra of (50-x)Na_2O-xZnO-50P_2O_5 glasses: (a) 0 mol% ZnO, (b) 5 mol% ZnO, (c) 10 mol% ZnO, (d) 15 mol% ZnO, (e) 20 mol% ZnO, (f) 25 mol% ZnO, (g) 30 mol% ZnO, (h) 33 mol% ZnO.
b. Infrared spectra of (0.9-x)$NaPO_3$-xSiO_2-0.1ZnO ($0 \leq x \leq 0.1$ mol) glasses: (a) 0 mol SiO_2, (b) 0.02 mol SiO_2, (c) 0.04 mol SiO_2, (d) 0.06 mol SiO_2, (e) 0.08 mol SiO_2, (f) 0.10 mol SiO_2.

Introducing alkali metal oxide or divalent metal oxide to the glass network induces the fundamental optical absorption edge falls in the UV region bellow 400 nm which meets with the requirement for desirable applications in optical systems [22]. These additions not only enhance the chemical durability of the phosphate glasses but also can impart special functions to the glasses and expand the glass application fields.

Furthermore, alkali phosphate glasses have attracted more attention due to their mixed electronic ionic conductivity, low melting point and strong glass-forming character [4]. Among the phosphate-based glasses those containing calcium, magnesium, sodium and zinc have received great attention due to their excellent optical properties, high refractive index, low dispersion and good transparency in the UV and IR region [1]. With the decrease of P_2O_5 content, the glass become more resistant to moisture attack but restricts the glass formation areas. Thus, MgO oxide was incorporated in order to overcome these problems [23].

Nevertheless, phosphate-based glasses containing transition metal ions are scientifically interesting materials due to their attractive properties which can be used in many technological applications including electronic and electro-optical devices [21].

In fact, transition metal oxides can be dissolved easily in phosphate glasses which exhibit one than more oxidation sates [21, 25, 30].

For glasses doped with manganese ions. These latter are presented in the +2 or +3 oxidation states. The content of Mn^{3+} ions in the glass leads to the staining glass in the color range from light to dark purple depending on the concentration [30]. This characteristic coloration could be explained by the d-d electronic transitions. This color can be associated with the broad absorption band in the visible region at 520 nm, which pertains to the Mn^{3+} ions [25, 30]. This behavior allows obtaining additional luminescence bands in the red spectral region that shift LED emission from cool white to warm light [30].

Contrary to Zn^{2+} and Mg^{2+} ions presented in one oxidation state, spectral-luminescent properties of manganese ions in phosphate glasses allow them to be good candidates for interesting optical applications [21, 25, 30].

In borate glasses, manganese exists mainly as Mn^{3+} ions with octahedral coordination in glass networks whereas in silicate and germinate glasses, it identified as Mn^{2+} ions with both tetrahedral and octahedral coordination [21].

Referring to literature, Montagne et al. have been studied the zinc phosphate glasses with a general formula (100-x)$NaPO_3$-xZnO with $0 < x < 33.3$ mol% using ^{31}P MAS-NMR, ^{31}P NMR of liquid sample, visible spectroscopy, refractive index measurements, density evolutions, T_g variations, activation energy, chemical durability and chemical analysis [1, 2, 46, 47].

The obtained results revealed the distortion of metaphosphate chains (Q^2) which suggests the formation of phosphate dimmers (Q^1) [46, 47].

Moreover, Zotov et al. have been studied the manganese phosphate glasses with a general formula $(MnO)_x(NaPO_3)_{1-x}$ when x = 0.0, 0.024, 0.048 and 0.167 mol [48].

These investigations have been performed using X ray diffraction, EXAFS and Raman spectroscopy. The increase of MnO content causes the depolymerization of metaphosphate chains leading to the decrease of the average chain length [48].

For zinc phosphate-silicate glasses, chemical compositions of the prepared glasses, picked from literature, have been chosen with higher level of SiO_2 and lower P_2O_5 content [2, 20, 24, 34].

Aguiar et al. have been studied the Na_2O-MgO-CaO-P_2O_5-SiO_2 bioactive glasses using Raman, ^{31}P MAS-NMR and ^{29}Si MAS-NMR spectroscopies. The glass compositions were prepared with varying SiO_2 content from 25 to 54 mol% and the P_2O_5 proportion from 2 to 11 mol% [2, 20, 24].

Furthermore, Szumera et al. have detailed the effect of MoO_3 addition on silicate phosphate glasses using spectroscopic analysis such as FTIR, Raman and ^{31}P MAS-NMR. The molar content of SiO_2 decreases from 41.6 to 39.6 mol% and the P_2O_5 proportion increases from 5.7 to 7.8 mol% [44]. The obtained results revealed the cleavage of oxygen bridges which suggests that acts as a network modifier [44].

In our knowledge, the thermochemical data of zinc-, magnesium-, and manganese-based phosphate and phosphate-silicate glasses are rare in literature. For this purpose, the aims of this research were to study the correlations between structural changes, thermal investigations, optical properties and thermochemical behaviors of these glassy compounds with the incorporation of zinc, magnesium and manganese oxides.

However, it is well known that phosphate glasses have a poor chemical durability, volatile nature and hygroscopic character. These disadvantages decrease their stability which limits their use in technological applications [5].

The addition of alkali and alkaline earth cations with the decease of phosphorus content can depolymerize the glass network which suggests the cleavage of P—O—P bridges [28].

The incorporation of certain network modifier cations (Na^+, K^+, Mg^+, Ca^{2+}...) disrupts the glassy network, leading to the structure depolymerization and the formation of non-bridging oxygen atoms (NBO), also named terminal oxygen (O_T) [2, 4].

The addition of transition metal oxides (CuO, MgO, ZnO, MnO, CaO...) into the vitreous network (TMO) disrupted the P—O—P bridges leading to the structure depolymerization and the formation of non-bridging oxygen atoms (NBO) [2, 5, 6, 8, 9, 13, 21, 25, 32, 35]. Which induces the formation of P—O—M bonds replacing the easily hydrosoluble P—O—P bridge that improve the chemical durability of the phosphate network [21].

Among several oxides mentioned above, zinc oxide gained considerable attention because Zinc doped glasses find numerous applications in optic field can be used as LED light sources and substrates for optical waveguides. It can also play an important role in bone formation and mineralization [3].

Zinc phosphate compositions are chemically durable, have processing temperatures under 400°C and can be co-formed with high temperature under 400°C polymers to produce unusual organic/inorganic composites [19].

In recent years, there have also appeared some publications on the influence of the addition of ZnO on the structure of glasses with a mixed phosphate-silicate structure [7, 34].

Among the wide class of phosphate glasses, ZnO-based glasses have low glass transition temperature in the region of 280–380°C and significantly high chemical durability [14].

Furthermore, the addition of ZnO to the glass network is expected to improve the chemical stability of the structure. It can also ameliorate the electrical, optical and magnetic properties of glasses due to the appearance of P—O—Zn ionic bond which induces the increase in the compactness and the rigidity of the glass network [5, 13, 32, 35].

In fact, in crystalline solid compound, the structure was described by a repetitive arrangement of a large scale patterns contrary to amorphous structure which exhibits a short range order.

Additionally, ZnO is an intermediate oxide. It can act as former or modifier network depending on his content in phosphate network. When it occupies the tetrahedral sites by forming ZnO_4 structural units, ZnO oxide plays the role of glass former. But, when it occupies the octahedral sites coordinated, ZnO oxide acts as glass modifier [7, 25].

The literature concerning zinc in various mixed oxide compounds revealed that with the exception of few structures, zinc has tetrahedral oxygen ligancy and the zinc-oxygen distance varies only slightly [25, 34]. It was concluded that zinc had a coordination number of six in borate and silicate glasses [34].

Interestingly, the heat treatment of glasses at temperature higher than their glass transition (T_g) or crystallization temperature (T_c) improves the electrical conductivity of the glassy compound due to the "structural relaxation" of the glass network [36].

ZnO is widely used in glass production because it improves the glass quality by enhancing mechanical properties and chemical durability and by reducing the thermal expansion [7]. Zinc is a microelement that plays an important role in the bone formation and mineralization. Zinc containing glasses and glass ceramics has been developed for bone engineering applications [7]. The small size of Zn ion (0.74 Å) helps it to locate itself into smaller cavities of the network [37].

Moreover, MgO oxide is of interest from a biological viewpoint because Mg^{2+} is known to play a physiological role in positively influencing bone strength which can be substituted into apatites [13, 18].

The bioactive behavior of magnesium rich glasses is identified as their ability to react chemically with living tissues, forming with them mechanically strong and lasting bonds. These bone bondings are attributed to the formation of an apatite-like layer on the glass surface, with composition and structure equivalent to the mineral phase of bone.

In fact, bioactive glasses have received special attention due to their better bone bonding ability in vivo. Due to their good bioactive and tailorable degradation properties, these glasses can be used for various biomedical applications such as bone graft, filler, dental, implant coating.

Furthermore, by increasing the concentration of modifier oxides, electrical conductivity of the glass increases. This property was probably influenced by the structural changes resulting from the disruption of the glass network which affected the mobility of the cations and anions when the modifying oxide was progressively introduced [38].

Glasses containing transition metal oxides possess interesting electronic, optic and magnetic properties due to the ability to exist in more than one valence state. However, the electronic conduction of these glassy compounds is resulting from the electronic transfer of cation that exists in different valence sates [4, 38].

Compared with phosphate glasses, silicate glasses exhibit superior chemical resistance which makes them compatible with the fabrication process in the development of optical devices [3].

Silicate glasses are an attractive host matrix for transition metal ions due to their excellent optical and mechanical properties, good chemical durability, good chemical stability and low thermal expansion coefficient leading to strong thermal resistance [6]. Silicate glasses have many advantages rather than phosphate glasses. Silicate-based glasses are chemically durable, thermally stable and optically transparent at excitation and lasing wavelength. However, the higher viscosity of these glasses allows the glass to be formed without crystallization process. In addition, these amorphous materials are useful in optics as lenses or beam splitters in optical telecommunications, micro- and optoelectronics and in near IR-windows [6, 39, 40].

2. Experimental procedures

2.1 Glass preparation

Glasses of the $(50\text{-}x/2)Na_2O\text{-}xMO\text{-}(50\text{-}x/2)P_2O_5$ (M = Zn, Mn, Mg) ($0 \leq x \leq 33$ mol%), $(50\text{-}x)Na_2O\text{-}xMO\text{-}50P_2O_5$ (M = Zn, Mn) ($0 \leq x \leq 33$ mol%), and

(0.9-x)$NaPO_3$-xSiO_2-0.1ZnO ($0 \leq x \leq 0.1$ mol) compositions have been prepared using a melt quenching technique.

A series of glasses were prepared by varying the MO (M = Zn, Mn, Mg) content from 0 to 33 mol% using reagent grade compounds, NaH_2PO_4, $NH_4H_2PO_4$, MgO, ZnO, $MnCO_3$ with a high purity (99% purity), in the suitable proportions.

The mixture corresponding to the desired compositions was heated in platinum crucible at 400°C in order to evaporate water and start the condensation of phosphate groups. The temperature was then progressively increased to 750–900°C, depending on glass composition, and held constant for 30 min. The batch was finally quenched to room temperature under air atmosphere in order to produce vitreous compounds.

Using the same technique, phosphate-based silicate glasses with a general formula (0.9-x)$NaPO_3$-xSiO_2-0.1ZnO ($0 \leq x \leq 0.1$ mol) have been synthesized using reagent grade compounds, $NaH_2PO_4.H_2O$ ZnO and SiO_2 with a high purity (99% purity) with the desired compositions.

The mixture was then putted in platinum crucible at 400°C for 1 hour in order to eliminate residual water. The temperature was raised progressively to 1200°C for 30 min in order to homogenize the melting mixture. Finally, the batch was quenched to room temperature under air atmosphere in order to obtain glasses.

The amorphous state was confirmed by X-ray diffraction. All the products were annealed at about 20°C below their glass transition temperature for 2 hours in order to eliminate internal tensions and get a more homogenized sample.

2.2 ICP analysis

Phosphorus, sodium, magnesium, zinc, manganese and silica were analyzed by inductively coupled plasma atomic emission spectroscopy (Jobin Yvon Ultra C).

2.3 Physical properties

2.3.1 Density and molar volume

Density of glass is a strong function of its composition and its intrinsic property which shed light on the short range structure of the glassy material [4, 37]. This work presents a series of glasses with various amounts of modifier oxides. These modifies, depending on their polarity and size, tend to occupy the interstices within the network and form new bonds resulting in a change in the structure and properties of the glass [4, 37].

Glass density measurements have been determined using the standard Archimedes method using diethyl orthophthalate as immersion fluid. The relative error of these measurements is ±3%.

The molar volume of glasses has been calculated from the density ($V_m = M/\rho$) and the molar weight.

For (50-x/2)Na_2O-xMO-(50-x/2)P_2O_5 (M = Zn, Mn, Mg) where $3 \leq O/P \leq 3.49$ and (50-x)Na_2O-xMO-50P_2O_5 (M = Zn, Mn) with O/P = 3 ($0 \leq x \leq 33$ mol%) series glasses, the density increases gradually with the incorporation of MO oxide. The increase in density indicates that the MO oxide reticulate the vitreous network because P—O—M bond are more ionic than P—O—P [11–13, 15, 29, 32, 33, 42].

Table 2 shows that the molar volume deceases monotonically with the increase of ZnO content for phosphate glasses.

The decrease in the molar volume for (50-x/2)Na_2O-xMO-(50-x/2)P_2O_5 (M = Zn, Mn, Mg) where $3 \leq O/P \leq 3.49$ and (50-x)Na_2O-xMO-50P_2O_5 (M = Zn, Mn) with O/P = 3 ($0 \leq x \leq 33$ mol%) series glasses could be explained by the higher field ΔF (ΔF = Z/r^2; with z is the valence cation and r is the ionic radius) of M^{2+} compared to that of Na^+ [11–13, 15, 29, 32, 33, 42].

The decrease in the molar volume is extensively related to structural changes due to the incorporation of MO oxide that disrupted the average chain length of metaphosphate resulting from the following reaction:

$$P—O—P + MO \rightarrow (2PO)M \tag{1}$$

The variation of these properties is closely related to the structural modification when M^{2+} ion is progressively introduced.

The effect of composition on the density and molar volume for zinc phosphate-silicate glass, having a general formula: (0.9-x)$NaPO_3$-xSiO_2-0.1ZnO ($0 \leq x \leq 0.1$ mol), shows that the replacement of $NaPO_3$ by SiO_2 oxide induces a decrease of density as mentioned **Table 3**. This is due to the lower molecular weight of SiO_2 than that of $NaPO_3$ ($M_{SiO_2} = 60$ gmol^{-1}, $M_{NaPO_3} = 102$ gmol^{-1}) [11–13, 15, 29, 32, 33, 42].

As the same for the molar volume, this quantity decreases monotonically with the incorporation of SiO_2 oxide (**Table 3**). This variation indicates that SiO_2 oxide reticulates the vitreous network suggesting the increase in the rigidity of the structure.

Furthermore, the regular decrease in the molar volume is closely related to the nature of bending in the glass structure, because P—O—Si are more ionic than P—O—P bridges, suggesting the compactness of the vitreous network [11–13, 15, 29, 32, 33, 42].

2.3.2 DSC investigations

Generally, the glass transition phenomenon occurs due to the increasing viscosity of the overcooled liquids so T_g strongly depends on the polymerization ratio of the network [40].

Glass composition	X	Density (g cm^{-3})	V_m (cm^3 mol^{-1})	T_g (°C)	T_c (°C)	ΔT (°C)
(50-x/2)Na_2O-xZnO-(50-x/2) P_2O_5	0	2.43 ± 0.07	42.00 ± 1.30	280 ± 5	290 ± 5	10
	5	2.47 ± 0.07	41.00 ± 1.23	-	-	-
	10	2.62 ± 0.08	38.15 ± 1.14	285 ± 5	371 ± 5	86
	15	2.70 ± 0.08	36.63 ± 1.10	-	-	-
	20	2.75 ± 0.08	35.60 ± 1.10	287 ± 5	368 ± 5	81
	25	2.85 ± 0.09	34.00 ± 1.02	294 ± 5	439 ± 5	145
	30	2.93 ± 0.09	32.70 ± 1.00	306 ± 5	456 ± 5	150
	33	2.98 ± 0.09	32.00 ± 1.00	314 ± 5	446 ± 5	132

Table 2.
Density, molar volume, glass composition, glass transition temperature T_g, T_c, ΔT of (50-x/2)Na_2O-xZnO-(50-x/2)P_2O_5 ($0 \leq x \leq 33$ mol%) phosphate glasses.

Glass composition	x	Density (gcm^{-3})	V_m (cm^3 mol^{-1})	n	T_g (°C)	T_c (°C)	ΔT (°C)
(0.9-x)$NaPO_3$-xSiO_2-0.1 ZnO	0	2.60 ± 0.10	38.50 ± 1.20	1.44 ± 0.05	280 ± 5	371 ± 5	86
	0.02	2.58 ± 0.10	38.35 ± 1.20	1.45 ± 0.05	287 ± 5	400 ± 5	113
	0.04	2.57 ± 0.10	38.24 ± 1.20	1.46 ± 0.05	289 ± 5	427 ± 5	138
	0.06	2.55 ± 0.10	38.13 ± 1.10	1.47 ± 0.05	293 ± 5	439 ± 5	146
	0.08	2.55 ± 0.10	38.00 ± 1.10	1.48 ± 0.05	294 ± 5	475 ± 5	163
	0.1	2.53 ± 0.10	37.80 ± 1.10	1.49 ± 0.05	296 ± 5	466 ± 5	170

Table 3.
Density, molar volume, refractive index, glass composition, glass transition temperature T_g, T_c, ΔT of (0.9-x) $NaPO_3$-xSiO_2-0.1 ZnO ($0 \leq x \leq 0.1$ mol) glass series.

The glass transition temperatures were determined on 40–50 mg of samples using DSC-ATD Netzsch 404 PC with a 10°C/min heating rate (accuracy ±5°C).

With increasing MO content, glass transition temperature, T_g, increases linearly for all glass compositions as mentioned **Table 2**.

This behavior is undoubtedly corresponding to some changes in the nature of bonding in the structural network. This parameters is strictly related to the bond strength of the glass network which can be explained in terms of bond length (which is the charge divided by the square of the cation-oxygen distance) affected by the cation field strength resulting in a higher of T_g values [11–13, 15, 29, 32, 33, 42].

These variations indicate the progressive increase of the reticulation and the rigidity of the glass network by gathering the non-bridging oxygen atoms (NBO) with the increase of MO proportion. As a result the formation of P—O—M bonds suggesting the increase in the rigidity and the compactness of the structure that ameliorate the chemical durability of glasses.

A similar behavior has been observed for zinc phosphate-silicate glasses [11–13, 15, 29, 32, 33, 42].

According to Dietezel, the thermal stability of glasses (ΔT) can be expressed by the temperature difference between T_g and T_c, $\Delta T = T_c - T_g$, in which T_g and T_c are the glass transition and crystallization temperature. Increasing ΔT delays the nucleation process, indicating a better stability of the glass [29].

Inspecting these data, one can note that the undoped ZnO and SiO_2 oxide glass matrix has the lowest thermal stability indicating a tendency towards crystallization as shown **Tables 2** and **3**.

ΔT increases when SiO_2 oxide is progressively added, indicating a better stability of the glass. The larger value of ΔT, the stronger is the inhibition to nucleation and crystallization process as mentioned **Table 3** [12, 13, 15, 29, 32, 33].

2.3.3 Refractive index measurements

The ability to control the physical properties of glasses, e.g., the refractive index, by variation in glass composition suggests the feasibility of chemically controlling the materials according to the needs of a given application [3].

For glassy compounds, refractive index is a fundamental parameter that strongly relevant to optical devices performance and reliability in the basic elements in all optical instruments. Hence, a large number of researchers have carried out investigations to ascertain the relation between refractive index and glass composition [10].

This parameter is one of the fundamental properties of materials, because it is closely related to the electric polarizability of ions and the local field inside the material [3, 10, 11].

The carat eristic feature of phosphate glasses is the low value of the refractive index that is in the order of 1.49. The variation of this quantity for zinc phosphate-based silicate glasses is presented in **Table 4**. Inspecting these data, one can note that n increases from 1.44 to 1.49 when x rises from 0 to 10 mol% of SiO_2 oxide which suggests that the refractive index of glassy compounds depends essentially on the density of glass network [3, 10, 11].

In addition to density, many parameters can prevails the refractive index such as density, polarizability of the first neighbor ions coordinated with it (anion), coordination number of ion, electronic polarizability of the oxide ion and optical basicity [3, 10, 11]. The molar refractivity (R_m) was estimated from the refractive index and the molar volume (V_m) using the Lorenz-Lorenz Equation [3, 10, 11]:

$$R_m = \frac{(n^2 - 1)}{(n^2 + 2)} V_m \tag{2}$$

Glass composition	n	α_m (Å^3)	R_m (cm^3 mol^{-1})	E_{opt} (ev)	$(\alpha_m/V_m) \times 10^{-25}$
$0.9NaPO_3$-0.1ZnO	1.44	4.06	10.23	1.50	1.05
$0.88NaPO_3$-$0.02SiO_2$-0.1ZnO	1.45	4.07	10.30	1.70	1.06
$0.86NaPO_3$-$0.04SiO_2$-0.1ZnO	1.46	4.13	10.40	2.00	1.08
$0.84NaPO_3$-$0.06SiO_2$-0.1ZnO	1.47	4.15	10.50	2.25	1.09
$0.82NaPO_3$-$0.08SiO_2$-0.1ZnO	1.48	4.18	10.55	2.35	1.10
$0.8NaPO_3$-$0.1SiO_2$-0.1ZnO	1.49	4.20	10.60	2.35	1.11

Table 4.
Refractive index, molar refractivity (R_m), molar electronic polarizability (α_m) and band gap energy of (0.9-x) $NaPO_3$-$xSiO_2$-0.1ZnO ($0 \leq x \leq 0.1$ Mol) glass series.

The molar electronic polarizability α_m was calculated using the relation of Clasius-Mosotti as follows [3, 10, 11]:

$$\alpha_m = \frac{3}{4}\Pi N R_m \tag{3}$$

where the value of $\frac{3}{4}\prod N$ is known as the Lorentz function and N is the Avogadro number. **Table 4** reports the values of R_m and α_m. These parameters increase gradually with the incorporation of SiO_2 oxide.

Table 4 shows that R_m increases from 10.23 to 10.60 m^3 mol^{-1} and α_m are between 4.06 and 4.20 Å. These variations indicate that the refractive index as a function of both density and molar electric polarizability of glassy compounds [3, 10, 11].

In the present work, we found that the refractive index (n) depends on the ratio $(\frac{\alpha_m}{V_m})$. This quantity shows that the refractive index (n) of the studied glasses increases linearly versus $(\frac{\alpha_m}{V_m})$ ratio. This variation can be probably due to the electronic polarizability of oxide ions.

For Na_2O ionic-based glasses, the polarizability of oxygen ions has the smaller value ($\alpha_{O^{2-}}$ = 2.45 Å) compared to copper rich glasses [3, 10, 11].

Duffy et al. suggested that increasing the optical basicity $\left(\Lambda = 1.67\left(1 - \frac{1}{\alpha_{O^{2-}}}\right)\right)$ indicates an increase in the effective electronic density of the oxide ions and accordingly, increasing covalency in the oxygen-cation bonding [3, 10, 11].

The decrease in the molar volume for zinc-based phosphate-silicate glasses induces an increase in the rigidity and the compactness of the vitreous network, when SiO_2 oxide is progressively introduced, because Si—O—P bonds are more ionic that P—O—P.

2.4 Spectroscopic analysis

2.4.1 FTIR spectroscopy

Infrared spectra of the glass series have been recorded by Perkin-Elmer (FTIR 2000) spectrometer using KBr pellets in the frequency range 400–4000 cm^{-1} at room temperature. The samples were prepared by grinding about 9 mg of glass powder with 300 mg of spectroscopic grade dried KBr.

For undoped zinc phosphate glasses, $NaPO_3$, FTIR spectrum revealed an asymmetric and symmetric stretching vibration band of PO_2 groups in metaphosphate chains situated respectively at 1280 and 1150 cm^{-1}. The asymmetric and symmetric stretching vibration bands of PO_3 chain end groups situated respectively at 1100 and 1000 cm^{-1} [12, 13, 15, 29, 32, 33].

Furthermore, the asymmetric and symmetric stretching vibrations of P—O—P bands are around 880, 780 and 720 cm^{-1}. The deformation mode of P—O—(PO_4^{3-}) groups at 535 and 480 cm^{-1} [12, 13, 15, 29, 32, 33].

FTIR spectra of (50-x/2)Na_2O-xZnO-(50-x/2)P_2O_5 and (0.9-x)$NaPO_3$-xSiO_2-0.1ZnO glasses are shown in **Figures 1a** and **b**.

As MO oxide is introduced, the asymmetric band of PO_2 shifts from 1280 cm^{-1} shift to lower frequency as showed **Figures 1a** and **b** indicating the depolymerization of phosphate chains when x increases [12, 13, 15, 29, 32, 33].

For higher ZnO content, FTIR spectra revealed the displacement of the asymmetric stretching mode vibration of the P—O—P band from 880 to 920 cm^{-1} when x rises from 0 to 33 mol%. This result can be correlated to the increase in the covalence character of P—O—P bridges when monovalent cation Na^+ was replaced by divalent cation (such as Zn^{2+}) [12, 13, 15, 29, 32, 33].

It may be also attributed to the shortening of phosphate chain length due to the higher field strength and the size of the metallic cation, when the ratio O/P increases for (50-x/2)Na_2O-xZnO-(50-x/2)P_2O_5 [12, 13, 15, 29, 32, 33].

The FTIR spectra of $NaPO_3$ glasses revealed also two bands in the frequency range 780–720 cm^{-1} which are attributed to the presence of two P—O—P bridges in metaphosphate chains based on $(P_2O_6)^{2-}$ groups (**Figure 1a** and **b**) [12, 13, 15, 29, 32, 33]. However, it is interesting to note that for the series glasses containing 30 and 33 mol% ZnO FTIR spectra exhibit only a single band at 740 cm^{-1} assigned to the P—O—P linkage in pyrophosphate group $(P_2O_7)^{4-}$ (**Figure 1a**) [12, 13, 15, 29, 32, 33]. These spectral changes depend essentially on the glass composition.

On the other hand, this result could be explained by disruption of the infinite metaphosphate chains when MO oxide is gradually incorporated suggesting the depolymerization of the skeleton of $(P_2O_6^{2-})_\infty$ into short phosphate groups such as: $P_2O_7^{4-}$ and PO_4^{3-} [12, 13, 15, 29, 32, 33].

Similar FTIR spectra have been recorded for zinc phosphate-based silicate glasses, with a general formula: (0.9-x)$NaPO_3$-xSiO_2-0.1ZnO, ($0 \leq x \leq 0.1$ mol) as mentioned **Figure 1b**.

The FTIR spectra of zinc phosphate-based silicate glasses revealed the appearance of some bands assigned to phosphate-silicate glasses in the range 1000–1300 cm^{-1} as shown **Figure 1b**. The asymmetric stretching vibration bands of silicate and phosphate tetrahedron is located at 1100 cm^{-1}. It seems that a smaller band at 840 cm^{-1} is attributed to the symmetric stretching vibration of O—Si—O in metasilicate (Q^2) when SiO_4 tetrahedron shared two oxygen with their neighbor (Si—O—2NBO). The symmetric stretching vibration band Si—O—Si is about 1080 cm^{-1}. In addition, the bending vibration of Si—O—Si and O—Si—O bonds is around 460 cm^{-1} [12, 13, 15, 29, 32, 33].

When SiO_2 is incorporated, FTIR spectra revealed the displacement of the asymmetric stretching mode vibration of PO_2 band from 1280 to 1250 cm^{-1} when x increases from 0 to 10 mol%. This result can be probably due to the depolymerization of the infinite metaphosphate chains with the addition of SiO_2 oxide.

Furthermore, for 0.8 $NaPO_3$-0.1 SiO_2-0.1 ZnO glass composition, FTIR spectrum revealed the appearance of a only a single band at 760 cm^{-1} assigned to P—O—P bands in phosphate dimers ($P_2O_7^{4-}$) as mentioned **Figure 1b**. These spectral changes have been observed for magnesium and manganese phosphate glasses that could be attributed to the reduction of infinite phosphate groups ($P_2O_6^{2-}$) into shorter phosphate groups such as: $P_2O_7^{4-}$ and PO_4^{3-} [12, 13, 15, 29, 32, 33].

2.4.2 Raman spectroscopy

Raman spectroscopy is an adequate technique for the analysis of glass matrix structure. Raman bands are generally characteristics of structures involving chains of linked tetrahedral that may be found in crystalline, glassy phosphates and silicate because it can detect the local changes in the environment of Si—O—Si and P—O—P bonds [1, 2].

The Raman spectra were recorded on powder of glasses using a Labram HR800 micro Raman model operating in the 50–4000 cm^{-1} range at room temperature equipped with an internal He-Ne laser source (λ = 488 nm).

Figure 2a reported the Raman spectra of zinc phosphate glasses having a general formula (50-x/2)Na_2O-xZnO-(50-x/2)P_2O_5 ($0 \leq x \leq 33$ mol%) with an O/P ratio varies from 3 to 3.49.

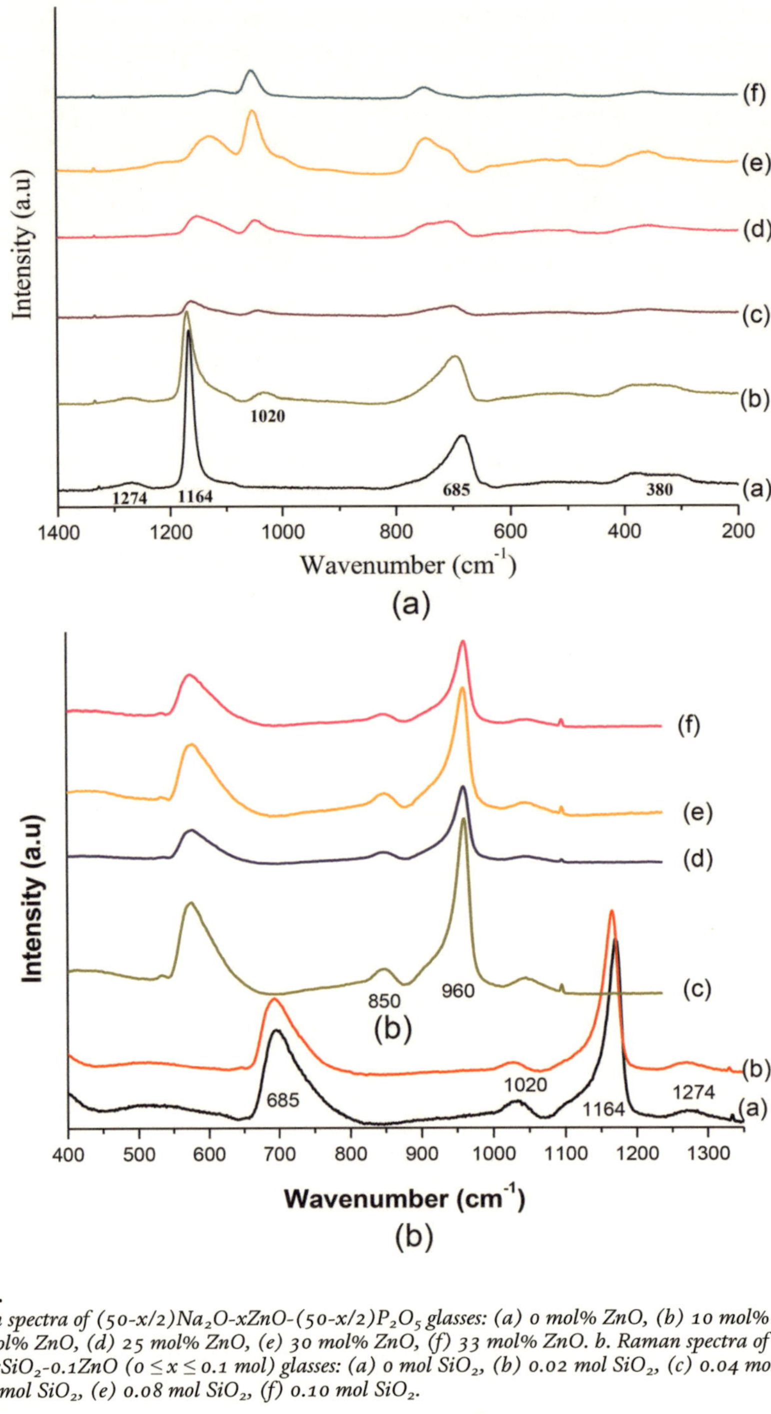

Figure 2.

a. Raman spectra of (50-x/2)Na_2O-xZnO-(50-x/2)P_2O_5 glasses: (a) 0 mol% ZnO, (b) 10 mol% ZnO, (c) 20 mol% ZnO, (d) 25 mol% ZnO, (e) 30 mol% ZnO, (f) 33 mol% ZnO. b. Raman spectra of (0.9-x) $NaPO_3$-xSiO_2-0.1ZnO ($0 \leq x \leq 0.1$ mol) glasses: (a) 0 mol SiO_2, (b) 0.02 mol SiO_2, (c) 0.04 mol SiO_2, (d) 0.06 mol SiO_2, (e) 0.08 mol SiO_2, (f) 0.10 mol SiO_2.

For undoped zinc phosphate glasses, Raman spectrum revealed a large band around 1274 cm^{-1} and three weaker peaks at 1164, 685, and 380 respectively as shown in **Figure 2a**.

The bands located at 1274 and 1164 cm^{-1} are assigned to the asymmetric and symmetric vibrations of PO_2 groups in metaphosphate chains (Q^2) [12, 13, 15, 29, 32, 33]. The large band at about 685 cm^{-1} is attributed to the symmetric vibration of the bridging oxygen linking two PO_4 tetrahedrons (P—O—P) in metaphosphate chains [12, 13, 15, 29, 32, 33]. The low frequency attributed to the faint band at 380 cm^{-1} is related to the bending motion of phosphate polyhedral [12, 13, 15, 29, 32, 33].

With increasing MO content, we observe some decrease of the overall background located at 600–800 cm^{-1} and 1100–1300 cm^{-1}. These spectral changes can be correlated to the distortion of P—O—P band which induces the shortening of the infinite metaphosphate chains suggesting the formation of pyrophosphate groups (Q^1) with the increase of the O/P ratio [12, 13, 15, 29, 32, 33].

From **Figure 2a**, it seems that the intensity of bands located at 1164 and 685 cm^{-1} decrease when MO oxide is progressively introduced. However, the Raman spectra revealed the displacement of these bands to higher frequencies to 1180 (d, e) and 780 cm^{-1} (d, e, f) with 30 and 33 mol% of MO level. This result can be probably due to the higher π character of P—NBO bands that induces the depolymerization of infinite metaphosphate chains when MO oxide is progressively added.

Similar Raman spectroscopic analysis have been recorded for (0.9-x) $NaPO_3$-$xSiO_2$-0.1ZnO ($0 \leq x \leq 0.1$ mol) glass compositions as shown in **Figure 2b**.

The incorporation of SiO_2 oxide to the phosphate network generates the appearance of the asymmetric band around 850 cm^{-1} attributed to Si—O—Si bending modes. The band located at 560 cm^{-1} is attributed to Si—O—Si intertetrahedral linkages obtained in calcium and magnesium rich silicate glasses in order to link the distorted metaphosphate groups when SiO_2 oxide is added [12, 13, 15, 29, 32, 33].

2.4.3 ^{31}P MAS-NMR spectroscopy

The ^{31}P MAS-NMR spectra of (50-x/2)Na_2O-xZnO-(50-x/2)P_2O_5 glasses are shown in **Figure 3**.

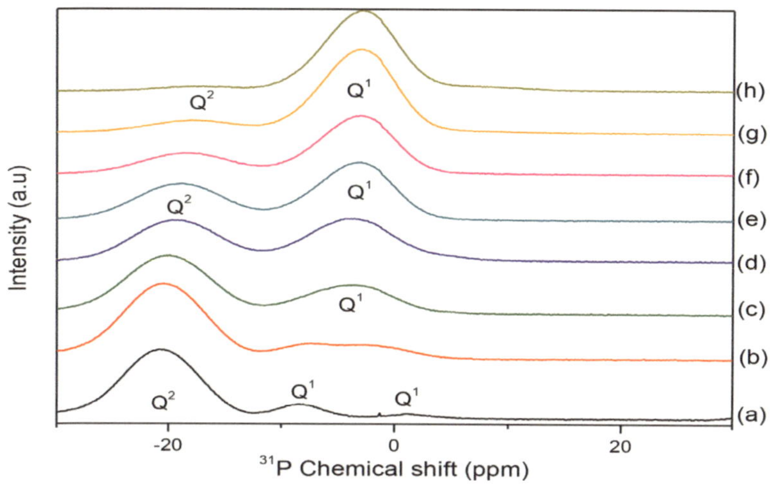

Figure 3.
^{31}P MAS-NMR spectra of (50-x/2)Na_2O-xZnO-(50-x/2)P_2O_5 glasses: (a) 0 mol% ZnO, (b) 5 mol% ZnO, (c) 10 mol% ZnO, (d) 15 mol% ZnO, (e) 20 mol% ZnO, (f) 25 mol% ZnO, (g) 30 mol% ZnO, (h) 33 mol% ZnO.

The characteristic features of undoped zinc phosphate glasses are isotopic peaks at −21 and −6.88 ppm. The first one is attributed to the Q^2 tetrahedral sites in metaphosphate groups and the second is assigned to the Q^1 groups at the end of chain [12, 13, 15, 29, 32, 33].

Based on literature, the chemical shift at +1.4 ppm is attributed to $NaPO_3$ chain end groups. From **Figure 3**, one can note the appearance of two isotopic peaks around 21––18.80 ppm and −6.88––3.90 ppm for the glass series.

When the MO oxide is introduced to the vitreous network, the intensity peak attributed to Q^1 tetrahedral sites increases and becomes the major spectral feature [12, 13, 15, 29, 32, 33]. These results are in good agreement for zinc phosphate glasses with the structural study of (100-x)$NaPO_3$-xZnO glasses ($0 \leq x \leq 33.3$ mol%) performed by Montagne et al. [12, 13, 15, 29, 32, 33].

From **Figure 3,** it seem that ^{31}P MAS-NMR spectra exhibit only single peak assigned to Q^1 tetrahedral sites attributed to pyrophosphate groups resulted from the distortion of metaphosphate chains when MO oxide is progressively introduced.

Furthermore, the phosphorus chemical shift depends essentially on the phosphorus-ligand bond (P-O) for phosphate compounds and the electronic density of the non-bridging oxygen (NBO). **Figure 3** mentioned that the Q^2 chemical shift becomes less shielded when MO is added. This decrease is probably due to the higher electronegativity of M^{2+} compared to Na^+ also the increase of π fraction of P-NBO resulting from the decondensation of phosphate chains when ZnO is incorporated. This suggests that Zn^{2+} ions are only bonded to pyrophosphate groups described by Q^1 tetrahedral sites. As a result the increase of shielding Q^2 sites from −21 ppm in $NaPO_3$ glass to −18.80 ppm in 33.5 Na_2O-33 ZnO-33.5 P_2O_5 glass composition [12, 13, 15, 29, 32, 33].

2.4.4 UV-VIS spectroscopy

UV-VIS-NIR absorption spectra of the glassy compounds were carried out by means of Perkin-Elmer Lambda 950 spectrometer at room temperature under air. Optical measurements were recorded in the range of 200 and 1800 nm.

Optical absorption, particularly the absorption edge, is useful for the investigation of optically-induced transitions and for getting information about the band gap energy [4, 10, 11, 14, 16, 37, 41]. This parameter is very interesting for the applications of the materials to be studied. It is known that the optical transition occurs through the region between conduction and valence bands (optical band gap) directly or indirectly.

However, the optical transition involves an energy transfer caused by electron transitions between conduction and valence bands [4, 10, 11, 14, 16, 37, 41].

The optical absorption coefficient α(hν) of the prepared glasses was calculated at different wavelengths by using the relation [4, 10, 11, 14, 16, 37, 41]:

$$\alpha = \frac{1}{d} \ln \left(\frac{I}{I_0} \right) \qquad (4)$$

where d represents the thickness of the glass composition and $\ln\left(\frac{I}{I_0}\right)$ is the absorbance.

For the optical measurements, one can note the absence of the absorption sharp edge which characterizes the vitreous nature of the prepared glasses [4, 10, 11, 14, 16, 37, 41].

According to Davis and Mott, the expression of the absorption coefficient α (ν) as a function of photon energy (hν) for direct and indirect optical absorption, was given by the relation as follows:

$$\alpha(\nu) = \frac{A\left(h\nu - E_{opt}\right)^n}{h\nu} \tag{5}$$

where

- A: an energy-independent constant
- E_{opt}: the optical band gap energy
- n: a constant which determines the type of the optical transition. For direct allowed transition $n = 2$ and in the case of indirect allowed transition $n = \frac{1}{2}$ [4, 10, 11, 14, 16, 37, 41].

For glassy materials, the indirect transitions are valid according to Tauc relations [4, 10, 11, 14, 16, 37, 41].

Figure 4 represents the variation $(\alpha h\nu)^2$ versus photon energy ($h\nu$) for (0.9-x) $NaPO_3$-xSiO_2-0.1ZnO ($0 \le x \le 0.1$ mol) series glasses.

The values of indirect optical band gap energy (E_{opt}) were determined by the extrapolation of the linear region of $(\alpha h\nu)^2$ against photon energy ($h\nu$) plots at $(\alpha h\nu)^2 = 0$. This latter shows that the E_{opt} increases with the incorporation SiO_2 from 1.5 to 2.35 eV. This quantity is not only influenced by the chemical composition also by the structural rearrangement in the glass matrix [4, 10, 11, 14, 16, 37, 41]. **Figure 5** shows clearly that the E_{opt} values dependent strongly on the composition of the glass also on the oxygen bonding in the vitreous network [4, 10, 11, 14, 16, 37, 41]. Any changes of oxygen bonding suggesting the formation of non-bridging oxygen (NBOs) causes a change of the absorption characteristics of the glass [4, 10, 11, 14, 16, 37, 41].

The higher energy is required to excite an electron from bridging oxygen (BO) than from non-bridging oxygen (NBO). As a result the increase in E_{opt} values [4, 10, 11, 14, 16, 37, 41].

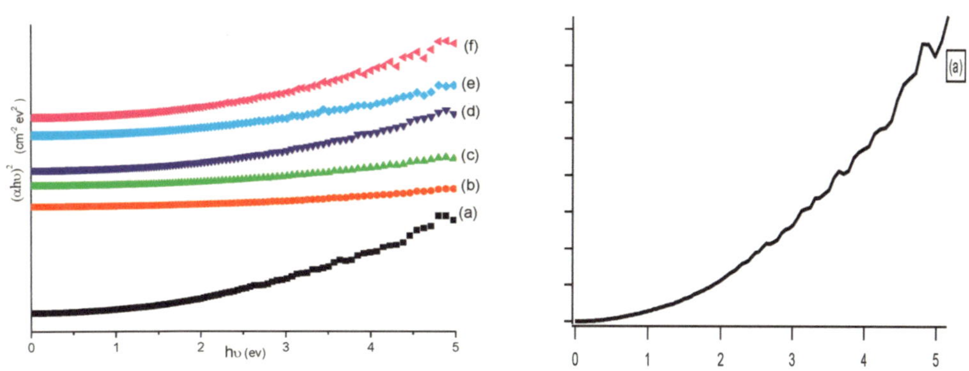

(a) The $(\alpha h\nu)^2$ as a function of photon energy of hv of $0.9NaPO_3$-0.1ZnO glass composition
*The E_{opt} values were determined by the extrapolation of the linear region of $(\alpha h\nu)^2$ against photon energy (hv) plots at $(\alpha h\nu)^2$=0.

Figure 4.
*The $(\alpha h\nu)^2$ as a function of photon energy of $h\nu$ of (0.9-x)$NaPO_3$-xSiO_2-0.1ZnO ($0 \le x \le 0.1$ mol) glasses: (a) 0 mol SiO_2, (b) 0.02 mol SiO_2, (c) 0.04 mol SiO_2, (d) 0.06 mol SiO_2, (e) 0.08 mol SiO_2, (f) 0.10 mol SiO_2. *Obtaining the lines corresponding to the curves of $(\alpha h\nu)^2$ against photon energy ($h\nu$) is probably due to the superposition effect for all the glass compositions.*

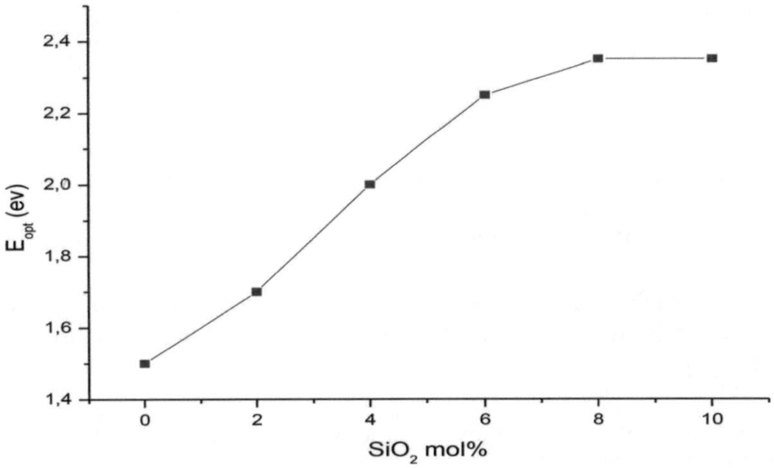

Figure 5.
Variation of optical band gap energy of (0.9-x)$NaPO_3$-xSiO_2-0.1ZnO ($0 \leq x \leq 0.1$ mol) glass series.

When x increases from 0 to 10 mol% of SiO_2, the optical band gap energy rises from 1.5 to 2.35 eV. This variation can be explained by the structural modifications which suggest the distortion of metaphosphate chains inducing the increase in the number of non-bridging oxygen (NBOs).

Because the NBOs bonds are predominantly ionic character and consequently have lower bond energies [34]. The higher value of the band gap energy revealed the increase of the cross-linking network due the introduction of SiO_2 [4, 10, 11, 14, 16, 37, 41].

From **Figure 5**, it seems that the E_{opt} is in the order of 2.35 eV for 0.82 $NaPO_3$-0.08 SiO_2-0.1 ZnO and 0.8 $NaPO_3$-0.1 SiO_2-0.1 ZnO glass compositions. This result can be correlated to the structural changes due to the formation of P-O-Si ionic bands [11].

3. Thermochemical study of phosphate glasses

3.1 Calorimetric dissolution of zinc, manganese and magnesium phosphate glasses

The calorimetric study was performed by determining the energy resulting from the dissolution of the glasses in a suitable solvent [11–13, 15, 29, 32, 33, 42].

Phosphate glasses are soluble in mineral acids [11–13, 15, 29, 32, 33, 42]. Furthermore, the dissolution process has been carried out in order to find the suitable solvent which dissolves entirely the glassy compounds and should not give rise to any secondary phenomena.

For this purpose, our investigations were covered all the usual acids, bases and their mixtures such as: HNO_3, HCl, NaOH, KOH, CH_3COOH.

The calorimetric profile shows that the 4.5% weight of phosphoric acid solution is the best solvent for the thermochemical requirements of phosphate glasses.

The dissolution of phosphate glasses were recorded by means the C80 (SETARAM) at 25°C. This equipment possesses two identical cells: the reference and the measuring cell. The reference cell should contain only the solvent but the measuring cell was provided with the solid to be dissolved or the liquid to be mixture. The superior compartment contains the attack solution (solvent) which is tightly separated from the lower one by a movable cover.

The reference and the measuring cell are surrounded by thermoelectric piles with high performance. These latters permit to detect the heat flow resulted from the dissolution, mixing or dilution process. The integration of the raw signal determined the heat dissolution of the studied compound.

Experiments were carried out by dissolving the same mass of solids (25 mg) in 4.5 ml of solvent.

The plots of heat dissolution of glasses ($\Delta_{sol}H$ (kJ mol^{-1})) are shown in **Figure 6**.

For (50-x/2)Na_2O-xZnO-50-x/2)P_2O_5 glass composition, it seems that the dissolution phenomenon is endothermic for lower ZnO content and becomes exothermic when ZnO oxide is progressively incorporated in the vitreous network.

Furthermore, the change in thermal signs is probably correlated to structural modifications of metaphosphate groups (Q^2) suggesting the formation of pyrophosphate units (Q^1) when ZnO oxide is introduced.

These results were correlated to spectroscopic investigations which revealed the formation of pyrophosphate groups for zinc, manganese and magnesium phosphate glasses, resulting from the cleavage of the P-O-P bridges when the amount of MO oxide is progressively increases [11–13, 15, 29, 32, 33, 42].

3.2 Thermochemical study of zinc phosphate glasses

Calorimetric study of glasses has been carried out for several decades. However, the thermochemical investigations of glassy compounds have been considered using a thermodynamic approach based on the Miedema's model in order to evaluate the formation enthalpy of binary alloys [42].

In the case of glassy compounds, the knowledge of the formation enthalpy is an important chemical data which can be used to determine the Gibbs free energy of formation of the selected compounds and to have an idea about their stability.

The determination of the formation enthalpy of (100-x)$NaPO_3$-xZnO glass series involves the formation enthalpy of sodium trimetaphosphate, $(NaPO_3)_3$, crystal). Because of the very old value reported in literature [42], this quantity has been determined again by the same technique.

3.2.1 Synthesis and characterization of samples

Sodium trimetaphosphate $(NaPO_3)_3$, was synthesized by thermal decomposition of sodium dihydrogen phosphate (NaH_2PO_4). This later was obtained by thermal

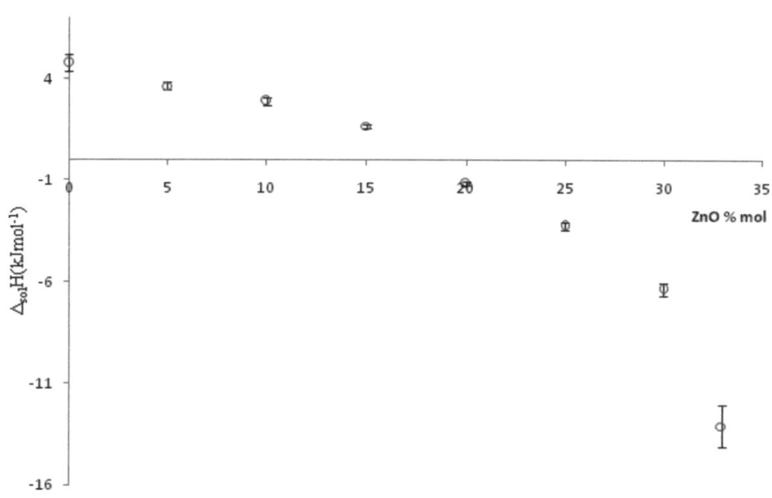

Figure 6.
Evolution of molar enthalpy of dissolution of (100-x)$NaPO_3$-xZnO($0 \leq x \leq 33$ mol%) glasses.

dehydration of its commercial monohydrate from $NaH_2PO_4.H_2O$ (Fluka of purity higher than 99%) at 150°C during 24 hours.

NaH_2PO_4 was placed in the furnace, then the temperature increases from 200 to 300°C during 24 hours in order to eliminate residual water and volatile gases. After any heat treatment, the powder was crushed.

Then, the temperature increases and maintains 500°C for one night. The final product was tested using an X-ray diffraction equipped by SEIFERT-XRD 3000 TT diffractometer which confirms that the final product is the sodium trimetaphosphate.

3.3 Cycle for sodium trimetaphosphate (($NaPO_3)_3$, crystal)

Generally, the direct determination of the formation enthalpy of any compounds is impossible. For this reason, our investigation is based on considering a particular reaction which involves the studied compounds with other reactants and products.

The knowledge of the enthalpy of the hypothetical reaction and the formation enthalpy of the reactants and products allow determining the formation enthalpy of the compound to be studied.

For the sodium trimetaphosphate (($NaPO_3)_3$, crystal); the following thermo-chemical cycle has been studied [42]. This later put into consideration the chemical reaction which involves dissolution, dilution and mixing processes.

The designed states from E1 to E4 are the solutions obtained from different chemical operations [42]:

- E1 state presents the phenomena of dissolution for NaH_2PO_4 in Slv2 (4.45%) ($\Delta_{sol}H_1$).
- E2 state designed the dissolution process of $(NaPO_3)_3$ in Slv1 (4.5%) ($\Delta_{sol}H_2$).
- E3 state designed the dilution process of $3H_2O_{(liq)}$ ($\Delta_{di}H_1$).
- E4 state designed the mixing process of E2 + E3 states ($\Delta_{mix}H_1$).

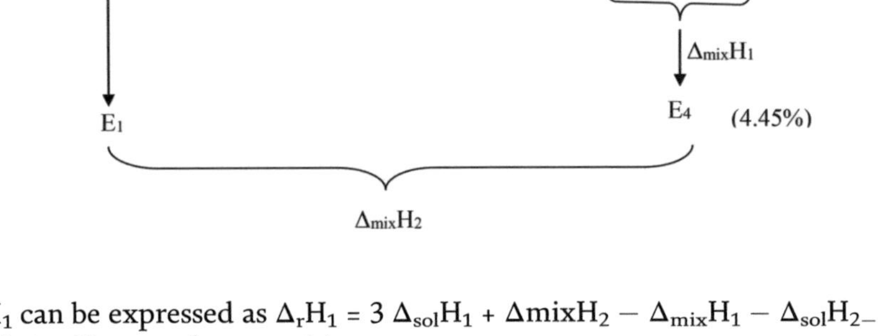

Δ_rH_1 can be expressed as $\Delta_rH_1 = 3\ \Delta_{sol}H_1 + \Delta mixH_2 - \Delta_{mix}H_1 - \Delta_{sol}H_{2-}\ \Delta_{dil}H_1$ in which $\Delta_{sol}H$ are molar quantities.

So the whole results allow to derive the enthalpy of reaction (R1) at 298.15 K. Taking into account the enthalpies of formation of NaH_2PO_4 (s) and H_2O (liq) [42], we can derive that for sodium trimetaphosphate ($(NaPO_3)_3$, (crystal)) at 298.15 K.

3.4 Cycle for Na_2O-ZnO-P_2O_5 series glass

For zinc-based phosphate glasses, the formation enthalpy of the glass series can be determined by considering a hypothetical reaction based on ($(NaPO_3)_3$(crystal)) and ZnO (sd). Their formation enthalpies can be derived by considering the following cycle which involves dissolution and mixing processes. The designed states from E5 to E8 are the solutions resulted from the different chemical operations [42]:

- E5 state presents the phenomena of dissolution for $(NaPO_3)_3$ in Slv1 (4.5%) ($\Delta_{sol}H_2$).
- E6 state designed the dissolution process of ZnO in Slv1 (4.5%) ($\Delta_{sol}H_3$).
- E7 state designed the mixing process of E5 + E6 states ($\Delta_{mix}H_3$).
- E8 state designed the dissolution process of glasses in Slv1 (4.5%) ($\Delta_{sol}H_4$).

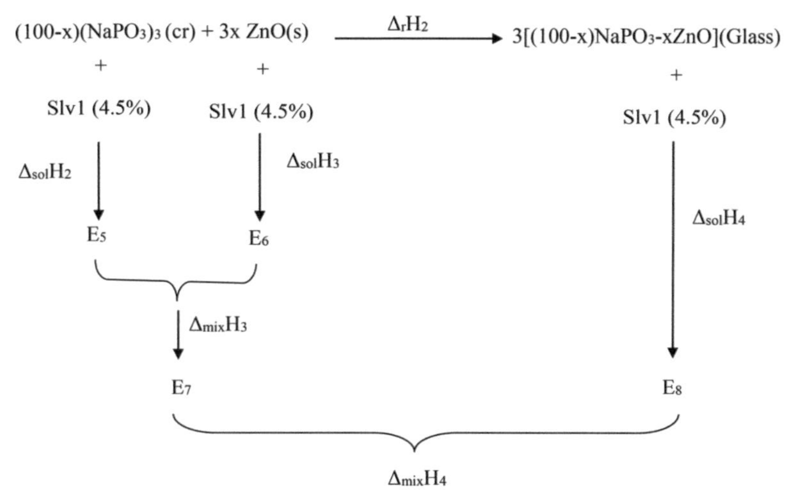

According to this cycle Δ_rH_2 can be expressed as: $\Delta_rH_2 = (100\text{-}x)\ \Delta_{sol}H_2 + 3x\ \Delta_{sol}H_3 + \Delta_{mix}H_3 + \Delta_{mix}H_4 - 3\Delta_{sol}H_4$ when E_7 and E_8 states are identical ($\Delta_{mix}H_4 \approx 0$) [42]. This quantity also equals:

$$\Delta_rH_2 = 3\ \Delta_fH\ [(100-x)NaPO_3 - xZnO, \text{glass}] - (100-x)\Delta_fH\ ((NaPO_3)_3, \text{crystal}) - 3x\ \Delta_fH\ (ZnO, \text{sd}) \quad (6)$$

So, the standard enthalpy of formation of the glass can be derived as:

$$\Delta_fH^\circ(\text{glass}) = 1/3(100-x)(\Delta_{sol}H(NaPO_3)_3, \text{cristal}) + \Delta_fH^\circ(NaPO_3)_3, \text{cr})) + x(\Delta_{sol}H\ (ZnO, \text{sd}) + \Delta_fH^\circ\ (ZnO, \text{sd})) - \Delta_{sol}H\ (\text{glass}) + 1/3\Delta_{mix}H_3 \quad (7)$$

3.4.1 Dissolution processes

Tables 5–7 show the dissolution heat (Q_r) of increasing the moles number (n) of NaH_2PO_4, $(NaPO_3)_3$ and ZnO solids in their corresponding solvents [42].

Table 5 presents the dissolution process of NaH_2PO_4 with the variation of the quantity (n (mmol)) to be dissolved in 4.5 ml of phosphoric acid solution (4.45% (w/w) H_3PO_4) (Slv2).

Table 6 presents the dissolution process of $(NaPO_3)_3$ with the variation of the quantity (n (mmol)) to be dissolved in 4.5 ml of phosphoric acid solution (4.5% (w/w) H_3PO_4) (Slv1).

Table 7 presents the dissolution process for ZnO with the variation of the quantity (n (mmol)) to be dissolved in 4.5 ml of phosphoric acid solution (4.5% (w/w) H_3PO_4) (Slv1).

The plots of the variation of the measuring heats as a function of the moles number of solid leads to straight lines whose expressed as: Q_r = An+b.

The slope (A) presents the molar dissolution enthalpy ($\Delta_{sol}H$) and b is the intercept increment.

Referring to a mathematical treatment developed in literature, the increment b is statistically not significant which leads to derive the dissolution enthalpy as [42]:

NaH_2PO_4 (sd) in Slv2 ($\Delta_{sol}H_1$)		
n (mmol)	**Q_r (J)**	**uQ_r (J)**
0.2548	1.084	0.05
0.3700	1.328	0.08
0.3440	1.282	0.06
0.3000	1.167	0.11
0.1366	0.680	0.06

Q_r: Heat of solution.
Slv2: 4.45% (w/w) H_3PO_4.
[a]Standard uncertainties u are u(T) = 0.01 K, u(p) = 10 kPa.

Table 5.
Enthalpy of solution of NaH_2PO_4 (sd) in 4.45% (w/w) H_3PO_4 at the temperature T = 298.15 K and pressure p = 0.1 MPa (level of confidence = 0.68).[a]

$(NaPO_3)_3$(cr) in Slv1 ($\Delta_{sol}H_2$)		
n (mmol)	**Q_r (J)**	**uQ_r (J)**
0.1995	0.950	0.08
0.2473	1.171	0.06
0.2990	1.370	0.07
0.3452	1.590	0.12
0.2626	1.240	0.10

Q_r: Heat of solution.
Slv1: 4.5% (w/w) H_3PO_4.
[a]Standard uncertainties u are u(T) = 0.01 K, u(p) = 10 kPa.

Table 6.
Heat of solution of $(NaPO_3)_3$(cr) in 4.5% (w/w) H_3PO_4 at the temperature T = 298.15 K and p = 0.1 MPa (level of confidence = 0.68).[a]

ZnO (sd) in Slv1 ($\Delta_{sol}H_3$)		
n (mmol)	**Q_r (J)**	**$u(Q_r)$ (J)**
0.2736	−26.081	0.12
0.3625	−34.600	0.33
0.2461	−23.500	0.24
0.2211	−21.215	0.22
0.3060	−29.124	0.50
0.3350	−32.010	0.21

Q_r: Heat of solution.
Slv1: 4.5% (w/w) H_3PO_4.
[a]Standard uncertainties u are u(T) = 0.01 K, u(p) = 10 kPa.

Table 7.
Enthalpy of solution of ZnO (sd) in 4.5% (w/w) H_3PO_4 at T = 298.15 K and p = 0.1 MPa (level of confidence = 0.68).[a]

$$\Delta_{sol}H = \frac{\sum(w_i * \Delta H_i n_i)}{\sum(w_i * n_i^2)} = A\ (\text{kJmol}^{-1})$$

where (w_i) is the reciprocal of the variance on ΔH_i ($w_i = 1/\sigma^2 \Delta H_i$), and ΔH_i is the energy resulting by dissolving n_i (mol) of the corresponding product in the phosphoric acid solution.

Equations of the lines are as follows [42]:

$$\Delta_{sol}H_1 = 4.01\ n \tag{8}$$

for (NaH_2PO_4) (sd) in 4.45% weight of H_3PO_4 solution.

$$\Delta_{sol}H_2 = 4.66\ n \tag{9}$$

for (($(NaPO_3)_3$, (cristal)) in 4.5% weight of H_3PO_4 solution.

For ZnO (sd) in 4.5% weight of H_3PO_4 solution.

$$\Delta_{sol}H_3 = -95.50\ n. \tag{10}$$

Table 8 gathers the values of molar dissolution enthalpies with the corresponding errors.

For zinc phosphate glasses, experiments were carried out by dissolving the same mass of solids (25 mg) in 4.5 ml of 4.5% weight of H_3PO_4 solution [42].

Compound	NaH_2PO_4 (sd) in Slv2	$(NaPO_3)_3$ (sd) in Slv1	ZnO (sd) in Slv1
$\Delta_{sol}H$ (kJ/mol)	$\Delta_{sol}H_1 = 4.01 \pm 0.47$	$\Delta_{sol}H_2 = 4.66 \pm 0.44$	$\Delta_{sol}H_3 = -95.5 \pm 2.7$

Slv1: 4.5% (w/w) H_3PO_4.
Slv2: 4.45% (w/w) H_3PO_4.
[a]Standard uncertainties u are u(T) = 0.01 K, u(p) = 10 kPa.

Table 8.
Molar enthalpy of solution of dissolved compounds at the temperature T = 298.15 K and pressure p = 0.1 MPa (level of confidence = 0.68).[a]

Composition	$\Delta_{sol}H$ (kJ/mol)	u ($\Delta_{sol}H$) (kJ/mol)
$NaPO_3$	4.80	0.45
95 $NaPO_3$-5 ZnO	3.70	0.20
90 $NaPO_3$-10 ZnO	2.92	0.15
85 $NaPO_3$-15 ZnO	1.70	0.10
80 $NaPO_3$-20 ZnO	−1.15	0.10
75 $NaPO_3$-25 ZnO	−3.21	0.20
70 $NaPO_3$-30 ZnO	−6.34	0.32
67 $NaPO_3$-33 ZnO	−13.05	1.00

[a]*Standard uncertainties u are u(T) = 0.01 K, u(p) = 10 kPa.*

Table 9.
Evolution of heat solution of (100-x)$NaPO_3$-xZnO phosphate glasses in 4.5% (w/w) H_3PO_4 at the temperature T = 298.15 K and pressure p = 0.1 MPa (level of confidence = 0.68).[a]

Table 9 reports the variation of the heat dissolution for the glass series published in a previous work [33]. This evolution shows that the dissolution heat decreases linearly with the incorporation of ZnO oxide. As a result the inversion in the thermal signs for the studies glasses. This variation can be explained by the structural changes of the phosphate network suggesting the distortion of metaphosphate chains revealed by ^{31}P MAS-NMR analysis.

Plotting of dissolution heat versus ZnO proportion ($\Delta_{sol}H$ (kJ mol^{-1})) is reported in **Figure 6**. It seems that the calorimetric dissolution of the glass series is endothermic for lower ZnO proportion and becomes exothermic above 18 mol% of ZnO. This variation can be correlated to the cleavage of P—O—P bridges which suggests the appearance of pyrophosphate groups (Q^1), revealed by ^{31}P MAS-NMR spectroscopic analysis, when ZnO oxide is progressively incorporated in the vitreous network [33].

3.4.2 Mixing processes

3.4.2.1 For (($NaPO_3$)$_3$, crystal) cycle

E2 was provided with various amounts of $(NaPO_3)_3$ (6–9 mg) which have been dissolved in 4.5% (w/w) H_3PO_4 solution but E3 is 4.4% (w/w) H_3PO_4 solution. Mixing the same volumes of E2 and E3 (around 2 ml) leads to a solution having the mean value of acid composition (E4 with 4.45% (w/w) H_3PO_4 or [H_3PO_4.116.90H_2O]. Consequently, the concentration of Slv2 was fixed as 4.45% (w/w) H_3PO_4 in order to get identical E1 and E4 states. **Table 10** reports E2 + E3 mixing enthalpy for different mole number (n) of $(NaPO_3)_3$ added in E2. This allowed to express $\Delta_{mix}H_1$ as: $\Delta_{mix}H_1 = 0.07$ n ($R^2 = 0.995$) leading to a value of 0.07 kJ per $(NaPO_3)_3$ mole [42].

($E_1 + E_4$) mixing process, which noted as ($\Delta_{mix}H_2$), was considered in order to check whether or not they correspond to the same final state. This operation led to an undetectable thermal effect [42].

3.4.2.2 For Na_2O-ZnO-P_2O_5 glasses cycle

Mixing the same volumes (around 2 ml) of E_5 and E_6 which have the same concentration of phosphoric acid (4.5% (w/w) H_3PO_4), led to E_7 solution. Were previously added to while The E_5 sate was obtained by dissolving the average mass of (($NaPO_3$)$_3$, (cristal)) (27.3 mg) whereas the E_6 state was considered by dissolving

n (mmol)	Q_r (J)	$u(Q_r)$ (J)
0.0187	0.0005	0
0.0257	0.0010	0
0.0157	0.0003	0
0.0300	0.0013	0

[a]*Standard uncertainties u are u(T) = 0.01 K, u(p) = 10 kPa.*
E2: Solution with various amounts of $(NaPO_3)_3$ *in 4.5% (w/w)* H_3PO_4.
E3: 4.4% (w/w) H_3PO_4.
Q_r*: Heat of mixing.*

Table 10.
Enthalpy of mixing: $\Delta_{mix}H_1$ *(E2 + E3) at the temperature T = 298.15 K and pressure p = 0.1 MPa (level of confidence = 0.68).*[a]

Composition	$\Delta_{mix}H_3$ (E5 + E6) (kJ/ZnO mole)	$u(\Delta_{mix}H_3)$ (kJ/mol)	$\Delta_{mix}H_4$ (E7 + E8) (kJ/mol)
95 $NaPO_3$-5 ZnO	10.15	0.10	≈0
90 $NaPO_3$-10 ZnO	4.00	0.13	≈0
85 $NaPO_3$-15 ZnO	5.10	0.21	≈0
80 $NaPO_3$-20 ZnO	3.23	0.24	≈0
75 $NaPO_3$-25 ZnO	2.74	0.25	≈0
70 $NaPO_3$-30 ZnO	3.00	0.15	≈0
67 $NaPO_3$-33 ZnO	2.12	0.20	≈0

[a]*Standard uncertainties u are u(T) = 0.01 K, u(p) = 10 kPa.*
E5: solution with 27.3 mg of $(NaPO_3)_3$ *in 4.5% (w/w)* H_3PO_4.
E6: solution with various amounts of ZnO in 4.5% (w/w) H_3PO_4.
E8: solution with 25 mg of glass composition in 4.5% (w/w) H_3PO_4.

Table 11.
Enthalpy of mixing: $\Delta_{mix}H_3$ *and* $\Delta_{mix}H_4$ *at T = 298.15 K and p = 0.1 MPa (level of confidence = 0.68).*[a]

the various amounts of ZnO so the variation of energy (**Table 11**) is due to the presence of ZnO in the solution. The mixture of E_7 and E_8 states has no detectable thermal effect [42].

3.4.3 Dilution processes

For the cycle corresponding to the sodium trimetaphosphate, addition of water to Slv1 [$H_3PO_4.115.54\ H_2O$] corresponds to a dilution process. The corresponding energy was calculated by linear interpolation of literature data considering the interval to which belongs each enthalpies of solution of the initial (4.5% (w/w) H_3PO_4) and final states (4.4% (w/w) H_3PO_4) [42].

The formation and dilution enthalpies were calculated from Ref. [42] and listed below:

$$\Delta_f H^\circ([H_3PO_4.115.54H_2O]) = -1288.255\ (H_3PO_4\ 4.5\%) \tag{11}$$

$$\Delta_f H^\circ([H_3PO_4.118.54H_2O]) = -1288.272\ (H_3PO_4\ 4.4\%) \tag{12}$$

Calculation gives $\Delta_{dil}H_1 = -0.017\ kJ\ mol^{-1}\ H_3PO_4$.

The formation enthalpy of sodium trimetaphosphate ($(NaPO_3)_3$, crystal) has been deduced as $(-3762.5 \pm 175)\ kJ\ mol^{-1}$. The obtained value differs from the

Composition	$\Delta_f H^\circ$(kJ/mol)	$u(\Delta_f H^\circ)$ (kJ/mol)
$NaPO_3$	−1260	50
95 $NaPO_3$-5 ZnO	−1213	49
90 $NaPO_3$-10 ZnO	−1174	47
85 $NaPO_3$-15 ZnO	−1132	45
80 $NaPO_3$-20 ZnO	−1090	44
75 $NaPO_3$-25 ZnO	−1070	43
70 $NaPO_3$-30 ZnO	−1003	40
67 $NaPO_3$-33 ZnO	−973	39

[a]*Standard uncertainties u are u(T) = 0.01 K, u(p) = 10 kPa.*

Table 12.
Formation enthalpy of: (100-x)$NaPO_3$-xZnO glasses at the temperature T = 298.15 K and pressure p = 0.1 MPa (level of confidence = 0.68).[a]

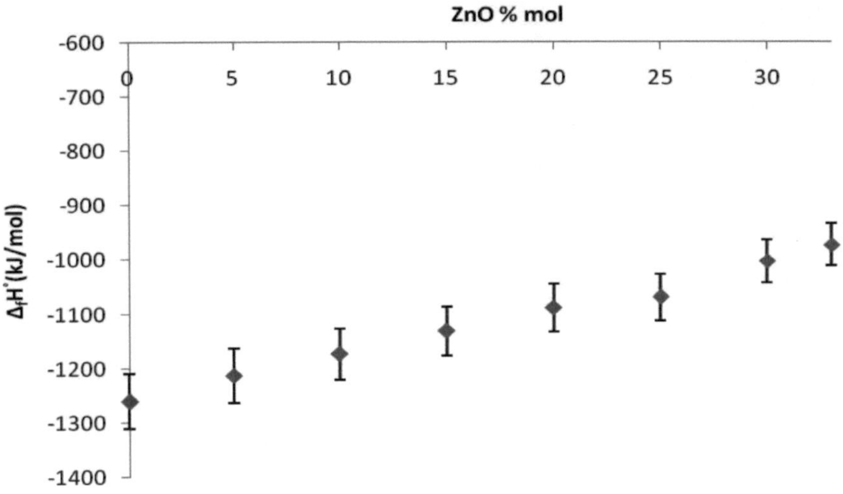

Figure 7.
Evolution of the standard enthalpy of formation at standard temperature and pressure of (100-x)$NaPO_3$-xZnO ($0 \leq x \leq 33$ mol%) glasses [45].

older by only 2.4%. It seems that the calculated value of the formation enthalpy of sodium trimetaphosphate ($(NaPO_3)_3$, crystal) is in good agreement with this determined previously in 1968. This confirms that the synthesized product is probably the sodium trimetaphosphate and not a mixture.

The variation of the formation enthalpy of the glass series are mentioned in **Table 12**. This latter shows that this quantity increases with the addition of ZnO oxide as reported in **Figure** 7 [42].

4. Conclusions

The influence of ZnO, MgO, MnO and SiO_2 addition on the structure, physical and optical properties of phosphate glasses and phosphate-based silicate glasses having a general formula: (50-x/2)Na_2O-xMO-(50-x/2)P_2O_5 (M = Zn, Mn, Mg) where $3 \leq O/P \leq 3.49$; (50-x)Na_2O-xMO-50P_2O_5 (M = Zn, Mn) with O/P = 3 ($0 \leq x \leq 33$ mol%) and (0.9-x)$NaPO_3$-xSiO_2-0.1ZnO ($0 \leq x \leq 0.1$ mol).

Amorphous state was investigated by means of FTIR, Raman, MAS-NMR and UV-visible spectroscopy in order to study the structural role of MO oxide.

Spectroscopic analysis revealed the formation of pyrophosphate groups (Q^1) resulting from the depolymerization of infinite metaphosphate groups (Q^2) when the modifying oxide is gradually incorporated.

Furthermore, the indirect optical band gap energy for zinc phosphate-based silicate glasses increases with the addition of SiO_2 oxide. This suggests the increase in the NBOs resulting from the modification of P—O—P bridges which revealed the shortening of the metaphosphate chains.

On the other hand, the dissolution process is endothermic at lower MO content and become exothermic when MO oxide is progressively incorporated. The change in thermal sign could be correlated to the structural modification inducing the formation of P—O—M ionic bond which increases the rigidity and the compacity of the vitreous network.

Furthermore, the glass formation enthalpy of (100-x)$NaPO_3$-xZnO ($0 \leq x \leq 33$ mol%) glass series were determine by considering a thermochemical cycle involving the formation enthalpy of sodium trimetaphosphate ($(NaPO_3)_3$, crystal). This later was checked in this work.

The glass formation enthalpy increases when ZnO oxide is progressively incorporated in the vitreous network.

When replacing of P by Zn induced a decrease in the binding energy which suggest the increase of the formation enthalpy of the glass series.

Furthermore, the variation of T_g values reflects an increase of the rigidity of the glass network due to the formation of P—O—Zn ionic bonds. As a result, the increase in the stability of the phosphate network which is tightly related to the Gibbs free energy of formation $\Delta_r G°$.

Because of the large disorder that exists in the vitreous structure, the entropy factor ($\Delta_r S°$) should prevail and induce the decrease in $\Delta_r G°$ value when ZnO concentration increases.

Author details

Refka Oueslati Omrani[1*], Mohamed Jemal[2], Ismail Khattech[2] and Ahmed Hichem Hamzaoui[1]

1 Useful Materials Valorization Laboratory, National Center For Research in Materials Sciences, Technological Park of Borj Cedria, Soliman, Tunisia

2 Université de Tunis El Manar, Faculty of Science, Chemistry Department, Laboratory of Materials Crystal and Applied Thermodynamics LR15SE01, Tunis, Tunisia

*Address all correspondence to: refkaoueslati@gmail.com

References

[1] Sreedhar VB, Basavapoornima CH, Jayasankar CK. Spectroscopic and fluorescence properties properties of Sm^{3+} doped zincfluorophosphate glasses. Journal of Rare Earths. 2014; **32**:918

[2] Aguiar H, Solla EL, Serra J, Conzalez P, Leon B, Malz F, et al. Jager, Raman and NMR study of bioactive Na_2O-MgO-CaO-P_2O_5-SiO_2 glasses. Journal of Non-Crystalline Solids. 2008; **354**:5004-5008

[3] Zid MHM, Matori KA, Hj S, Aziz A, Zakaria A. Effect of ZnO on the physical and optical band gap of soda lime silicate glass. International Journal of Molecular Sciences. 2012;**13**: 7550-7558

[4] Ibrahim AM, Bader AM, Elshaikh HA, Mostafa AG, Elbashar YH. Effect of CuO addition on the dielectric parameters of sodium zinc phosphate glasses. Silicon. 2017. pp. 1265-1274

[5] Vedeanu N, Stanescu R, Filip S, Ardelean I, Cozar O. IR and ESR investigations on V_2O_5-P_2O_5-BaO glass system with opto-electronic potential. Journal of Non-Crystalline Solids. 2012; **358**:1881-1885

[6] Ahmina W, El Moudane M, Zriouil M, Taibi M. Effect of the content of MnO on the electric-dielectric properties of potassium-phosphate glasses. Journal of Materials and Environmental Science. 2017;**11**: 4193-4198

[7] Waclawska I, Szumera M, Sulowska J. Structural characterization of zinc-modified glasses from the SiO_2-K_2O-CaO-MgO. Journal of Alloys and Compounds. 2016;**666**:352-358

[8] Pascuta P, Bosca M, Borodi G, Vida-Simiti I, Culea E. Thermal, structural and magnetic properties of some zinc phosphate glasses doped with manganese ions. Journal of Alloys and Compounds. 2011;**509**:4314-4319

[9] Oui MA, Azooz MA, Elbatal HA. Optical and infrared spectral investigations of cadmium zinc phosphate glasses doped with WO_3 or MoO_3 before and after subjecting to gamma irradiation. Journal of Non-Crystalline Solids. 2018;**494**:31-39

[10] Khor SF, Talib A, Mat Yunus WM. Optical properties of ternary zinc magnesium phosphate glasses. Ceramics International. 2012;**38**:935-940

[11] Omrani RO, Hamzaoui AH, Chtourou R, M'nif A. Structural, thermal and optical properties of phosphate glasses doped with SiO_2. Journal of Non-Crystalline Solids. 2018; **481**:10-16

[12] Cherbib MA, Khattech I, Montagne L, Reval B, Jemal M. Structure properties relationship in calcium sodium metaphosphate and polyphosphate glasses. Journal of Non-Crystalline Solids. 2018;**485**:1-13

[13] Oueslati-Omrani R, Kaoutar A, El Jazouli A, Krimi S, Khattech I, Jemal M, et al. Structural and thermochemical properties of sodium magnesium phosphate glasses. Journal of Alloys and Compounds. 2015;**632**:766-771

[14] El-Maaref AA, Badr S, ElOkr KS, Abdel Wahab EA, ElOkr MM. Optical properties and radiatives rates of Nd^{3+} doped zinc-sodium phosphate glasses. Journal of Rare Earths. 2019;**37**:253-259

[15] Cherbib MA, Krimi S, El Jazouli A, Khattech I, Montagne L, Reval B, et al. Structure and thermochemical study of strontium sodium phosphate glasses. Journal of Non-Crystalline Solids. 2016; **447**:59-65

[16] Magdas DA, Stefan R, Toloman D, Vedeanu NS. Copper ions influence on lead-phosphate glass network. Journal of Molecular Structure. 2014;**1056-1057**: 314-318

[17] Vedeanu N, Magdas DA, Stefan R. Structural modifications induced by addition of copper oxide to lead-phosphate glasses. Journal of Non-Crystalline Solids. 2012;**358**: 3170-3174

[18] Zhang L, Liu S. Structure and crystallization behavior of 50CuO-$xTiO_2$-(50-x)P_2O_5. Journal of Non-Crystalline Solids. 2017;**473**:108-113

[19] Brow RK. Review: The structure simple of phosphate glasses. Journal of Non-Crystalline Solids. 2000;**263-264**: 1-28

[20] Aguiar H, Solla JM, Serra J, Gonzalez P, Leon B, Almeida N, et al. Orthophosphate nanostructures in SiO_2-P_2O_5-CaO-Na_2O-MgO biactive glasses. Journal of Non-Crystalline Solids. 2008; **354**:4075-4080

[21] Abd el Ghany HA. Physicl and optical characterization of manganese ions in sodium-zinc-phosphate glass matrix. IARJSET. 2018. p. 5

[22] Khor SF, Talib ZA, Malek F, Cheng EM. Optical properties of ultraphosphate glasses containing mixed divalent zinc and magnesium ions. Optical Materials. 2013;**35**:629-633

[23] Walter G, Vogel J, Hoppe U, Hartmann P. The structure of CaO-Na_2O-MgO-P_2O_5 invert glass. Journal of Non-Crystalline Solids. 2001;**296**: 212-223

[24] Li HC, Wang DG, Hu JH, Chen CZ. Influence of fluoride additions on biological and mechanical properties of Na_2O-CaO-SiO_2-P_2O_5 glass-ceramics. Materials Letters. 2013;**106**:373-376

[25] Pascuta P, Borodi G, Jumate N, Vida-Simiti I, Viorel D, Culea E. The structural role of manganese ions in some zinc phosphate glasses and glass ceramics. Journal of Alloys and Compounds. 2010;**504**:479-483

[26] Pop L, Bolundut L, Pascuta P, Culea E. Influence of Er^{3+} ions addition on thermal and optical properties of phosphate-germanate system. Journal of Thermal Analysis and Calorimetry. 2019

[27] Jlassi I, Elhouichet H, Ferid M. Electrical conductivity and dielectric properties of MgO doped lithium phosphate glasses. Journal of Physics E. 2016;**81**:219-223

[28] Ciceo-Lucacel R, Todea M, Simon V. Effect of selenium addition on network connectivity in P_2O_5-CaO-MgO-Na_2O glasses. Journal of Non-Crystalline Solids. 2018;**480**:10-13

[29] Cherbib MA, Khattech I, Montagne L, Reval B, Jemal M. Effect of SrO content on the structure and properties of sodium-strontium metaphosphate glasses. The Journal of Physical Chemistry A. 2017:62-68

[30] Sevastiajova I, Aseev V, Tuzova L, Fedorov Y, Nikonorov N. Spectral and luminescence properties of manganese ions in vitreous lead metaphosphate. Journal of Luminescence. 2019:495-499

[31] Ciceo-Lucacel R, Todea M, Simon V. Effect of selenium addition on network connectivity in P_2O_5-CaO-MgO-Na_2O glasses. Journal of Non-Crystalline Solids. 2018;**488**:10-13

[32] Oueslati-Omrani R, Krimi S, Videau JJ, Khattech I, El Jazouli A, Jemal M. Structural investigations and calorimetric dissolution of manganese phosphate glasses. Journal of Crystalline Solids. 2014;**389**:66-71

[33] Oueslati-Omrani R, Krimi S, Videau JJ, Khattech I, El Jazouli A,

Jemal M. Structural and thermochemical study of Na_2O-ZnO-P_2O_5 glasses. Journal of Crystalline Solids. 2014;**390**: 5-12

[34] Hurt JC, Phillips JC. Structural role of zinc oxide in glasses in the system Na_2O-ZnO-SiO_2. Journal of the American Ceramic Society. 1967;**53**

[35] Wan MH, Wong PS, Hussin R, Lintang HO, Edud S. Structural and luminescence properties of Mn^{2+} ions doped calcium zinc borophosphate glasses. Journal of Alloys and Compounds. 2014;**595**:39-45

[36] Hassaan MY, Moustafa MG, Osouda K, Kubuki S, Nishiba T. 57Fe and 119Sn Mössbauer, XRD, FTIR and DC conductivity study of Li_2O-Fe_2O_3-SnO_2-P_2O_5 glass and glass ceramics. Journal of Alloys and Compounds. 2018; **765**:121-127

[37] Mohan S, Kaur S, Kaur P, Singh DP. Spectroscopic investigations of Sm^{3+} doped lead alumino-borate glasses containing zinc, lithium and barium oxides. Journal of Alloys and Compounds. 2018;**763**:486-495

[38] Maji BK, Jena H, Asuvathraman R. Electric conductivity and glass transition temperature (Tg) measurements on some selected glasses used for nuclear waste immobilization. Journal of Non-Crystalline Solids. 2016; **434**:102-107

[39] Berwal N, Dhankhar S, Sharma P, Kundu RS, Punia R, Kishore N. Physical, structural and optical characterization of silicate modified bismuth-borate-tellurite glsses. Journal of Molecular Structure. 2017;**1127**:636-644

[40] Belova EV, Kolyagin YA, Uspenskaya IA. Structure and glass transition temperature of sodium-silicate glasses doped with iron. Journal of Non-Crystalline Solids. 2015; **423-424**:50-57

[41] Hammad AH, Abdel-Hameed SAM, Margha FH. Effects of crystallization and microstructure on the dc electrical conductivity in the system xCuO-(70-x) MnO-30SiO_2. Journal of Alloys and Compounds. 2015;**627**:423-429

[42] Oueslati-Omrani R, Khattech I, Jemal M. Standard formation enthalpy of Na_2O-ZnO-P_2O_5 series glasses. Chemistry Africa. 2018;**1**:43-51

[43] Cimek J, Stepien R, Klimczak M, Zalewska I. Developpement of thermally stable glass from SiO_2-Bi_2O_3-PbO-ZnO-BaO oxide system suitable for all solid photonic crystal fibers. Optical Materials. 2017;**73**:277-283

[44] Jlassi I, Sdiri N, Elhouichet H, Ferid M. Raman and impedance spectroscopy methods of P_2O_5-Li_2O-Al_2O_3 system doped with MgO. Journal of Alloys and Compounds. 2016;**645**: 125-130

[45] Videau JJ, Flem L. Les verres phosphates de la spécificité de l'atome de phosphore à la formation, la structure, et la durabilité chimiques de phosphates vitreux. Institut de chimie de la matière condensée de Bordeaux; 27 October 2010. HAL Id: Cel-00530128

[46] Montagne L, Palavit G, Dalaval R. 31P NMR in (100-x)$NaPO_3$-xZnO glasses. Journal of Non-Crystalline Solids. 1997;**215**:1-10

[47] Montagne L, Palavit G, Delaval R. Effect of ZnO on the properties of (100-x)$NaPO_3$-xZnO glasses. Journal of Non-Crystalline Solids. 1998;**223**:43-47

[48] Zotov N, Schlenz H, Brendebach B, Modrow H, Hormes J, Reinauer F, et al. Effects of MnO doping on the structure of sodium metaphosphate glasses. 2003; **58a**:419-428

Chapter 10

Temperature Influence on Inhibitory Efficiency of Three Phosphate Inhibitors by Mass Loss

Latefa Sail

Abstract

The effect of temperature on steel samples immersed in concrete pore solutions contaminated by chlorides incorporating three inhibitors based on phosphate (Na_3PO_4, K_2HPO_4, and Na_2PO_3F) was studied by gravimetric measurements at several ranges: 298, 308, and 318 K. The results obtained for the use of these three products show that the inhibitory efficacy is lower at 318 K than that detected at 308 and 298 K of temperature. Also, we find that the best inhibitory efficiency at 298 K was detected for Na_2PO_3F (75.80% at 0.05 mol/l of concentration) followed by K_2HPO_4 (65.05% at 2.5 10^{-3} mol/l) and then Na_3PO_4 (61.48% at 7.5 10^{-3} mol/l).

Keywords: temperature, concrete pores, corrosion inhibitors, phosphate, gravimetric measurement efficiency

1. Introduction

Corrosion of reinforcement in concrete is one of the most dangerous pathologies that attack reinforced concrete structures; the means of protection against corrosion are varied and expensive. During this last decade, a new alternative has been adapted which is the application of corrosion inhibitors either as an adjunct to the mass of fresh concrete or by impregnation on the facing of hardened concrete. Several families of corrosion inhibitor products have been developed to prove their protective effect against steel reinforcement corrosion initiated by the penetration of chlorides through the pores of concrete. The best known are phosphates, borates, silicates and carbonates. One of the peculiarities of these ions is that their hydrolysis releases hydroxide ions which will have the effect of increasing the pH of the medium and thus passivating the steel. Moreover, in the presence of oxygen, the anions will form with the metal cation a very insoluble iron III phosphate which will clog the anodic surface and displace the cathodic reduction reaction [1].

The required concentration of passivative inhibitor, often of the order of 10^{-4} to 10^{-5} mol/l [2], it depends in fact on many factors such as temperature, pH, the presence of depassivating ions such as chlorides or reducing agents such as sulfur S_2 [3].

Temperature is one of the factors that can alter the behavior of a material in a corrosive environment. It can modify the metal-inhibitory interaction in a medium [4].

The variation of temperature affects the rate of corrosion. According to Liu and Weyer [5], an increase in temperature increases the rate of corrosion. This result was confirmed in carbonated concrete and also that subject to aggressive environments like chloride ions penetration.

The objective of this research is based on the analysis of the evolution of the inhibitory efficiencies of three phosphate inhibitors (Na_3PO_4, K_2HPO_4 and Na_2PO_3F) as a function of the temperature variation: 298, 303 and 313 K.

2. Methods and measurements

In this section, gravimetric tests were performed to characterize the influence of temperature on inhibition efficiency for the three phosphate inhibitors used in this study.

2.1 Gravimetric measurements

These measurements consist in determining the weight loss of a steel sample subjected to specified conditions of temperature and relative humidity; they are calculated on the basis of three tests to determine the average. The steel sample is polished with abrasive paper ranging from 120 up to 1000 grades using a polisher at a speed of 500 rpm, then rinsed in distilled water, dried with an electric dryer then we weigh the mass M1.

The steel samples are introduced into beakers containing 50 ml of electrolytic solution in an inclined position as shown in **Figure 1**, hermetically closed, then they are placed in a thermostatic bath while adjusting the desired temperature, after 24 h, the samples are removed from beakers then, rinsed in distilled water, degreasing is carried out with acetone and then dried with the electric dryer, after that we weigh the mass M2.

2.2 Study medium

The medium of this study is a concrete synthetic medium which simulates concrete pores contaminated by 3% of chlorides given in **Table 1**.

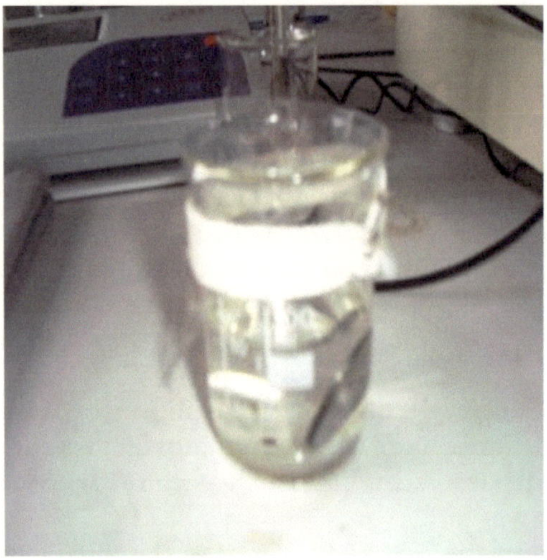

Figure 1.
Position of steel sample.

1 l distilled water	$Ca(OH)_2$	NaOH	KOH	$CaSO_4 2H_2O$	NaCl
Wt (g/l)	2	0.4	0.56	0.27	30

Table 1.
Synthetic medium of concrete [6–8].

2.3 Steel preparation

The steel used is circular shaped with a diameter of 27 ± 1 and 2 ± 2 mm of thickness, and the procedure of gravimetric tests is detailed in [9]. The corrosion rate is determined by the following formula:

$$Cr = \Delta M/S.t \left(mg/h.cm^2\right) \quad (1)$$

$$\Delta M = M1 - M2 \quad (2)$$

Hence, ΔM represents the difference between the initial mass M1 and the final mass M2 after a time "t" equal to immersion time by hours. "S" is the surface of the metal exposed to the electrolytic solution.

This value of the corrosion rate is the average of three tests carried out under the same conditions for an optimal concentration at a definite time. The value of the inhibitory efficiency is given by the following formula:

$$IE\ (\%) = \frac{Cr0 - Cr}{C_{r0}}.100 \quad (3)$$

3. Tested inhibitors

This study describes the corrosion behavior of steel immersed in synthetic concrete pore solutions contaminated by chlorides for three phosphate-inhibitors (Na_3PO_4, K_2HPO_4 and Na_2PO_3F), their molecular structure is given

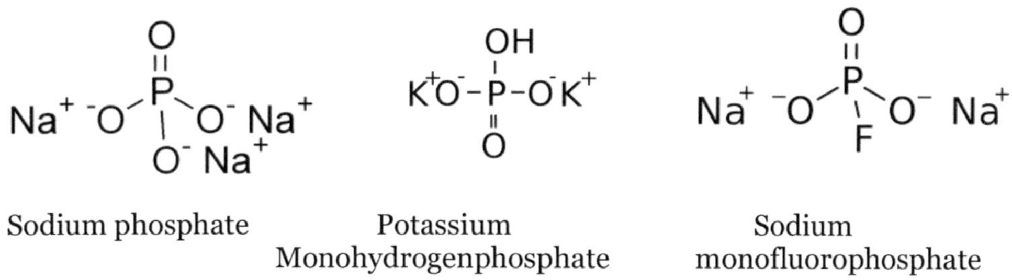

Figure 2.
Molecular structure of the three tested inhibitors.

Inhibitor	Concentration (mol/l)
Na_3PO_4	7.5×10^{-3}
K_2HPO_4	2.5×10^{-3}
Na_2PO_3F	5×10^{-2}

Table 2.
Medium concentrations.

in **Figure 2**. The optimal concentration which provides maximum efficiencies for the three products cited was extracted from a previous study [10] (see **Table 2**).

4. Results and discussions

Table 3 records the mass loss results, relating to the evolution of corrosion rates as well as the inhibitory efficiencies as a function of the temperature variation: 298, 303 and 318 K for the three inhibitors.

It can be seen from the results shown in **Table 3** that the corrosion rates decrease in the presence of the corrosion inhibitor, it reached the maximum at the optimal concentration, for the first inhibitor sodium phosphate Na_3PO_4 the maximum efficiency 69.28% was detected at a concentration of 7.5×10^{-3} mol/l at 298 K, we can see clearly that the inhibitory efficiency slightly decrease as a function of temperature increase. Likewise for K_2HPO_4, the best efficiency 67.44% was detected at 298 K for a concentration of 2.5×10^{-3} mol/l, also, the increase of temperature affects the inhibitory efficiency which decrease following temperature increasing, the same remark was recorded for Na_2PO_3F the maximal efficiency 75.8% was detected at 298 K. This phenomenon can be explained by the fact that the anodic processes (oxidation components of steel) and cathodic (proton reduction in acidic medium) are thermally activated.

This results in a current of exchange which increases the corrosion rate. Hunkeler [11] has shown in his studies that the influence of temperature on the rate of corrosion is greater than that the resistivity of the concrete.

Figure 3 shows the evolution of inhibitory efficiencies as a function of temperature variation for different concentrations of tested inhibitors.

	298 K		303 K		313 K	
C (mol/l)	Cr 10^{-3} (mg/h.cm^2)	IE %	Cr 10^{-3} (mg/h.cm^2)	IE %	Cr 10^{-3} (mg/h.cm^2)	IE %
Na_3PO_4						
0	1.4	—	2.5	—	3.1	—
5×10^{-3}	0.444	68.28	0.964	61.44	1.231	60.29
7.5×10^{-3}	0.43	69.28	0.798	68.08	1.0875	64.92
10^{-2}	0.768	45.14	1.49	40.4	1.8764	39.47
K_2HPO_4						
0	1.4	—	2.5	—	3.1	—
10^{-3}	0.4629	66.93	0.8889	64.44	1.2478	59.77
2.5×10^{-3}	0.4558	67.44	0.8737	65.04	1.1737	62.14
5×10^{-3}	0.4689	66.5	0.89126	64.34	1.251	59.45
Na_2PO_3F						
	1.4	—	2.5	—	3.1	—
2.5×10^{-2}	0.3411	75.63	0.8675	65.29	1.3976	54.91
5×10^{-2}	0.3386	75.8	0.6749	73	0.8986	71.01
7.5×10^{-2}	0.3414	75.61	0.8344	66.62	1.1842	61.8

Table 3.
Evolution of corrosion rates and inhibitory efficiencies as a function of temperature variation.

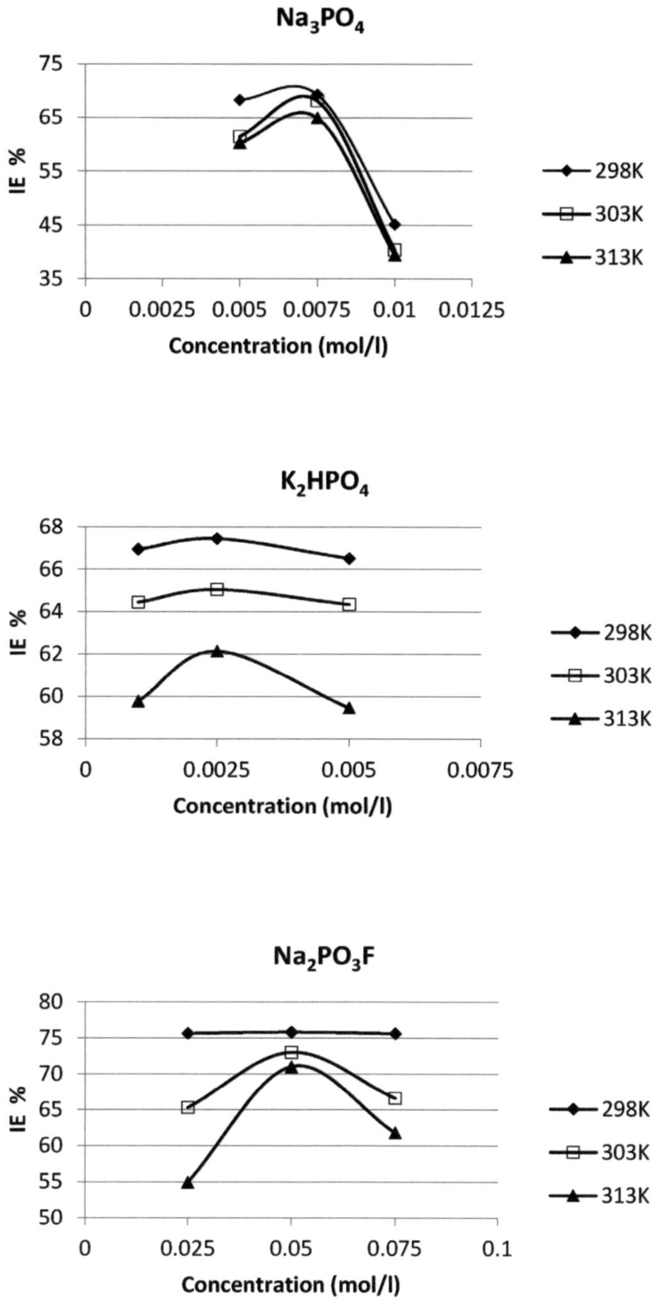

Figure 3.
Evolution of inhibitory efficiencies as a function of temperature variation for different concentrations of Na_3PO_4, K_2HPO_4 *and* Na_2PO_3F.

Figure 3 illustrates the influence of temperature variation on the inhibitory efficacy of the three phosphate-based inhibitors. Certainly, temperature is one of the factors that can alter the behavior of a material in a corrosive environment. It can modify the metal-inhibitory interaction in a given environment [4].

The increase in temperature causes the instability of inhibitory molecules and also reduces the inhibitory efficacies which was detected in previous researches [5].

It can be seen from **Figure 4** that inhibitory efficiencies are highest in the optimum concentration for all the studied temperature ranges, although they decrease

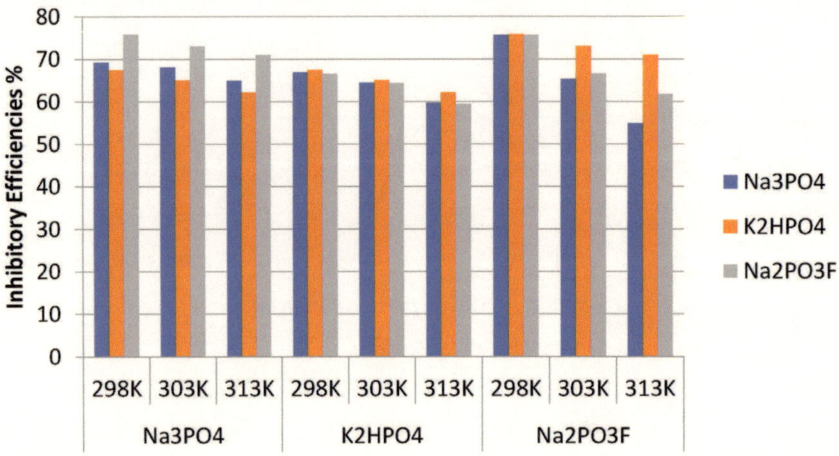

Figure 4.
Evolution of inhibitory efficiencies as a function of temperature variation.

slightly as a function of temperature increase. As a result, the maximum inhibitory efficacy at T 298, 303 and 318 K deduced using gravimetric measurements was confirmed by sodium monofluorophosphate (Na_2PO_3F), followed by potassium monohydrogenphosphate (K_2HPO_4) and thirdly sodium phosphate (Na_3PO_4).

These results are in good agreement with previous research that used the same inhibitory products [10].

Indeed, sodium monofluorophosphate has been the subject of several studies [12–14], and it has proven remarkable inhibitory properties especially in the case of its use in zinc phosphate baths [15–17].

The variation of the temperature influences the rate of corrosion and consequently the mechanism of the inhibition [18]. According to Liu and Weyer [5], an increase in temperature increases the rate of corrosion.

5. Conclusions

Direct measurements of both corrosion rates and inhibitory efficiencies as a function of inhibitor concentrations, have confirmed that sodium monofluorophosphate (Na_2PO_3F) offers the best corrosion protection under study conditions (temperature 298, 303 and 313 K); its inhibitory efficiency has exceeded 70% for these temperatures.

This inhibitor has been the subject of several previous studies [12, 19–22], its effectiveness against corrosion has been confirmed especially when used in a carbonated concrete [14, 23–26] and also than for concrete solutions contaminated by chlorides.

We can also conclude that increase of temperature affects inhibitory efficiencies, which is in good concordance with literature. For inhibitors based of phosphate, the increase of temperature has a slight influence on the inhibitory efficiency for the study temperatures, moreover, at higher temperatures, the molecular activation will be greater, which leads to an increase in corrosion rates.

Acknowledgements

My sincere gratitude and thanks go to the members of the granular materials team: corrosion prevention at the EOLE laboratory Department of Civil Engineering University of Tlemcen-Algeria.

Conflict of interest

My 'research areas' focus on corrosion prevention and repair by inhibitors.

Author details

Latefa Sail
Faculty of Technology, Aboubekr Belkaid University, Tlemcen, Algeria

*Address all correspondence to: saillatefa@yahoo.fr

References

[1] Oly M. Contribution à l'évaluation des capacités des glycérophosphates pour la maintenance dans le béton armé, [thesis]. France: University of Toulouse; 2011

[2] Buchler M, Corrosion inhibitors for reinforced concrete, in Corrosion Reinforced Concrete Structures. In: H. Böhni. Switzerland: Formerly Swiss Federal Institute of Technology; 2005. pp. 190-214

[3] Helie M. Matériaux métalliques—Phénomènes de corrosion, Edition Ellypses. Val of Essonne: University of Evry; 2015. ISBN: 9782340004023

[4] Khenadeki A. Etude théorique et expérimentale de l'effet d'inhibition de la corrosion d'un acier au carbone par les dérivées de base de Schiff en milieu acide chlorhydrique [thesis]. Algeria: University of Tlemcen; 2013

[5] Liu T, Weyer RW. Modeling the dynamic corrosion process in chloride contaminated concrete structures. Cement and Concrete Research. 1998;**28**(3):365-379. DOI: 10.1016/S0008-8846(98)00259-2

[6] Ghods P, Isgor OB, Mcrae B, Millar T. The effect of concrete pore solution composition on the quality of passive oxide films on black steel reinforcement. Cement and Concrete Composites. 2009. DOI: 10.1016/j.cemconcomp.2008.10.0032-11

[7] Moragues A, Macias A, Andrade C. Equilibria of the chemical composition of the concrete pore solution. Part I: Comparative study of synthetic and extracted solutions. Cement and Concrete Research. 1987;**17**(2):173-182. DOI: 10.1016/0008-8846(87)90100-1

[8] Page CL, Vennesland O. Pore solution compositions and chloride binding capacity of silica fume cement paste. Materials and Structures. 1983;**16**(1): 19-25. DOI: 10.1007/BF02474863.

[9] Sail L, Ghomari F, Bezzar A, Benali O. Mass loss for assessment of the inhibitory efficiency of products to basis of phosphate. Canadian Journal on Environmental, Construction and Civil Engineering. 2011;**2**(5):111-117. Availabe from: http://www.ampublisher.com/June%202011/ECCE-1106-018-Mass-loss-for-assessment-of-the-inhibitory.pdf

[10] Sail L. Etude de la performance d'inhibiteurs de corrosion à base de phosphate pour les constructions en béton armé [thesis]. Algeria: Tlemcen; 2013

[11] Hunkeler F. Grundlagen der korrosion und der potential messing baustahlbetonbauten, ASTRA Brücken unter halts for schung. Zürich: Verein Schweizer Strassen fachleute (VSS); Report No. 510; 1994

[12] Douche-Portanguen A, Prince W, Lutz T, Arliguie G. Detection or quantitative analysis of a corrosion inhibitor, the sodium monofluorophosphate, in concrete. Cement and Concrete Composites. 2005;**27**(6):679-687. DOI: 10.1016/j.cemconcomp.2004.11.002

[13] Pujol Lesueur VN. Etude du mécanisme d'action du monofluorophosphate de sodium comme inhibiteur de la corrosion des armatures métalliques dans le béton [thesis]. Paris, France: University of Pierre et Marie Curie; 2004

[14] Duprat M, Bonnel A, Dabisi F. Les monofluorophosphates de zinc et de potassium en tant qu'inhibiteurs de la corrosion d'un acier au carbone en solution de NaCl à 3%. Journal of Applied Electrochemistry. 1983;**13**:317-323

[15] Zimmermann D, Mun˜oz AG, Schule JW. Formation of Zn-Ni alloys in the phosphating of Zn layers. Surface & Coatings Technology. 2005;**197**(2): 260-269. DOI: 10.1016/j.surfcoat. 2004.07.129

[16] Kashyap A. Effects of water chemistry, temperature, gaseous cavitation et phosphate inhibitors on concrete corrosion [Master]. USA: University of Virginia; 2008

[17] Simescu F. Elaboration des revtements de phosphates de zinc sur armature à béton. Etude de leur comportement à la corrosion en milieu neutre et alcalin [thesis]. France: University of Lyon; 2008

[18] Khouikhi F. Etude de l'efficacité de deux inhibiteurs de corrosion dans les milieux multiphasiques (Eau, huile et gaz) [thesis]. Boumerdes, Algeria: University of M'Hamed Bougara; 2007

[19] Soylev TA, Richardson MG. Corrosion inhibitors for steel in concrete. State-of-the-art report. Construction and Building Materials. 2008;**22**(4):609-622. DOI: 10.1016/j. conbuildmat.2006.10.013

[20] Farcas F, Chaussadent T, Fiaud C, Mabille I. Determination of the sodium monofluorophosphates in a hardened cement paste by ion chromatography. Analytica Chimica Acta. 2002;**472**(1-2):37-43. DOI: 10.1016/S0003-2670(02)00978-9

[21] Talamge JM, Biemer TA. Determination of potassium nitrate and sodium monofluorophosphate in the presence of phosphate and sulphate by high resolution ion chromaltography. Journal of Chromatography. 1987;**410**:494-499. DOI: 10.1016/S0021-9673(00)90084-0

[22] Wang P, Li SF, Lee HK. Simultaneous determination of monofluorophosphate and fluoride in toothpaste by capillary electrophoresis. Journal of Chromatography. 1999. DOI: 10.1016/S0021-9673(96)00926-0

[23] Alonso C, Andrade C, Argiz C, Malric B. Na_2PO_3F as inhibitor of corroding reinforcement in carbonated concrete. Cement and Concrete Research. 1996;**26**(3):405-415. DOI: 10.1016/S0008-8846(96)85028-9

[24] Vézina D. Performance des inhibiteurs de corrosion, direction des laboratoires et chaussées. Technical Newsletter. 1997;**2**:3

[25] Benzina Mechmeche L, Dhouibi L, Ben Ouezdou M, Triki E, Zucchi F. Investigation of the early effectiveness of an amino-alcohol based corrosion inhibitor using simulated pore solutions and mortar specimens. Cement and Concrete Composites. 2008;**30**:167-173. DOI: 10.1016/j. cemconcomp.2007.05.007

[26] Dhouibi L, Triki E, Salta M, Rodrigues P, Raharinaivo A. Studies on corrosion inhibition of steel reinforcement by phosphate and nitrite. Materials and Structures. 2003;**36**(8):530-540. DOI: 10.1007/ BF02480830